Berichte aus der Automatisierungstechnik

Boris Lohmann / Axel Gräser (Hrsg.)

Methoden und Anwendungen der Automatisierungstechnik

Methods and Applications in Automation

24. Kolloquium der Automatisierungstechnik in Salzhausen

24th Colloquium of Automation, Salzhausen, Germany

Shaker Verlag
Aachen 2003

Bibliografische Information der Deutschen Bibliothek
Die Deutsche Bibliothek verzeichnet diese Publikation in der Deutschen Nationalbibliografie; detaillierte bibliografische Daten sind im Internet über http://dnb.ddb.de abrufbar.

Printed in Germany.

ISBN 3-8322-2117-4
ISSN 0945-4659

Shaker Verlag GmbH • Postfach 101818 • 52018 Aachen
Telefon: 02407 / 95 96 - 0 • Telefax: 02407 / 95 96 - 9
Internet: www.shaker.de • eMail: info@shaker.de

Vorwort

Das jährlich stattfindende *Kolloquium der Automatisierungstechnik, Salzhausen*, wird zunehmend von internationalen Gästen besucht, die in der Regel in englischer Sprache vortragen. Die Herausgeber des vorliegenden Sammelbandes ausgesuchter Beiträge der Jahre 2001 und 2002 möchten diesem Trend Rechnung tragen, und haben die Autoren im Sinne einer größtmöglichen Verbreitung und Akzeptanz ermutigt, *englischsprachige* Aufsätze beizutragen – mit großer Resonanz, wie ein Blick in das Inhaltsverzeichnis erkennen lässt. Wie schon in den Jahren zuvor finden sich Beiträge aus Theorie und Anwendung unterschiedlichster Gebiete der Automatisierungstechnik.

Für die engagierte Organisation der Kolloquien 2001/02 danken wir den Herren Bernd Augustin, Nikola Nedeljkovic, Sven Fischer und Bernd Hillers vom Institut für Automatisierungstechnik der Universität Bremen. Herrn Fischer gilt außerdem besonderer Dank für Durchführung der verschiedenen Arbeiten, die von der Planung bis zum Erscheinen dieses Bandes notwendig waren.

More and more international guests participate in the Colloquium of Automation in Salzhausen, Germany, which takes place every year. For this reason a big part of the talks are given in English. The editors of this anthology of selected contributions from the years 2001 and 2002 wanted to take this development into account. They encouraged the German authors to send in their papers in English in order to achieve the widest possible dissemination and acceptance of their results. The response was very good as a look at the table of contents will show you. Like in the years before you will find contributions about theory as well as about practical applications from different fields of automation.

We would like to thank Bernd Augustin, Nikola Nedeljkovic, Sven Fischer and Bernd Hillers from the Institute of Automation at the University of Bremen for having organized the colloquia in 2001 and 2002. Our special thanks go to Sven Fischer for his great effort during the publication process of this volume.

Bremen, im Mai 2003 Boris Lohmann Axel Gräser

Inhaltsverzeichnis

Theory

Image Processing and Robotics

Introduction to Krylov Subspace Methods in Model Order Reduction

Boris Lohmann and Behnam Salimbahrami,
Institut für Automatisierungstechnik, Universität Bremen, 28359 Bremen,
BL@iat.uni-bremen.de, Salimbahrami@iat.uni-bremen.de

In recent years, Krylov subspace methods have become popular tools for computing reduced order models of high order linear time invariant systems. The reduction can be done by applying a projection from high order to lower order space using the bases of some subspaces called input and output Krylov subspaces. The aim of this paper is to give an introduction into the principles of Krylov subspace based model reduction and to give a rough overview of the algorithms available for computation. A technical example illustrates the procedures.

1 Introduction and Principles of Model Order Reduction

Krylov subspace methods play a key role in the approximate solution of many large-scale problems in different scientific fields[1]. In model order reduction, they can be used to find a mapping from the high-dimensional space of a given state-space model to some lower dimensional space and vice versa, thereby defining a model of reduced order. This procedure of using mappings in order reduction is also referred to as *reduction by projection*. It turns out that also classical methods in linear order reduction, like modal methods [5] and balancing and truncation [14], can be understood as projection methods. Therefore, in this introductory section, we briefly review some basic concepts and relations in model order reduction.

For simplicity, we consider linear time-invariant state-space models with only one input and one output (although all methods discussed here can be extended to multi-input multi-output systems),

$$\dot{\mathbf{x}} = \mathbf{A}\mathbf{x} + \mathbf{b}u\,, \tag{1}$$

$$y = \mathbf{c}^T\mathbf{x}\,, \tag{2}$$

with the input variable u, the output variable y, the $(n,1)$-vector $\mathbf{x}$ of state variables, a constant non-singular matrix $\mathbf{A}$, and vectors $\mathbf{b}$, $\mathbf{c}$ of suitable dimensions. The number of state-variables, n, is called the *order* of the system. The transfer function of the system in (1),(2) is

$$g(s) = \mathbf{c}^{\mathrm{T}}(s\mathbf{I} - \mathbf{A})^{-1}\mathbf{b}\,, \tag{3}$$

relating the output to the input by $y(s) = g(s)u(s)$.

If the number of state variables, n, is very high, model analysis, simulation and design are difficult. Figures 1 and 2 illustrate two technical systems that can be modelled in state-space by equations of type (1) and (2), leading to system orders between 400 and several thousands (depending on how detailed modeling is done). When dealing with such high-order models, it is reasonable to look for models of lower order:

---

[1] Krylov subspace iteration methods were mentioned among the "top 10 algorithms of the 20th century" in IEEE Computing in Science and Engineering, Vol 2 (2000), Issue 1.

The task of *model order reduction* is finding a model (typically of type (1),(2)) with a significantly lower order, $q < n$, that approximates the behaviour of the original model.

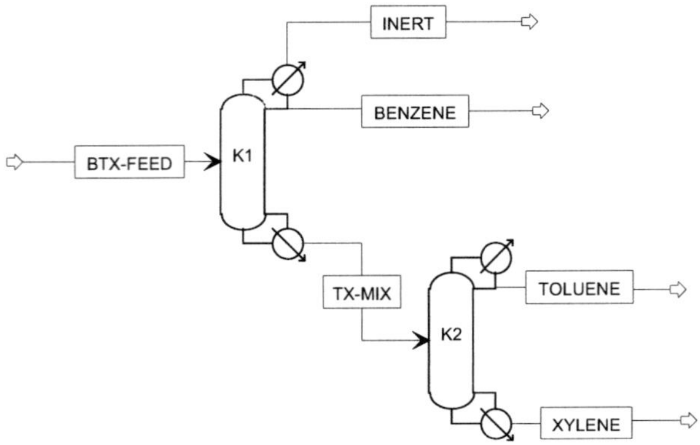

Figure 1 Block diagram of a destillation column (model by BASF AG). The BTX-Feed is a mixture of benzene, toluene and orhto-xylene, to be separated into three separate streams. A computer-aided modeling with linearisation at an operating point delivers $n = 400$ first-order ordinary differential equations

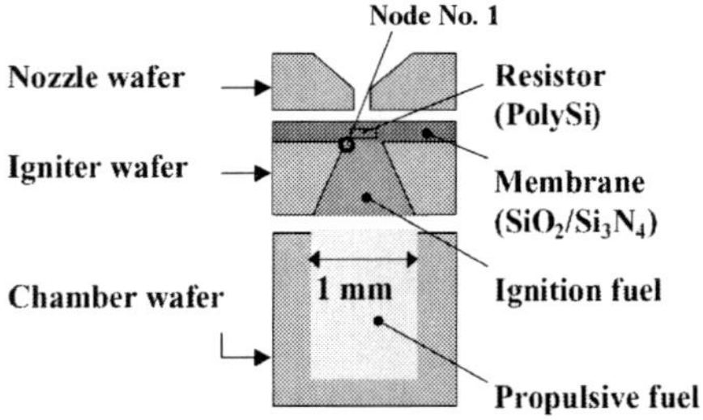

Figure 2 Components of a micro-fluidic system (IMTEK, Univ. Freiburg [3]). The resistor is heated up by electrical current until the ignition fuel and the propulsive fuel explode. The heating-up process can be modelled by finite elements methods and leads to 1000 to 10000 ordinary differential equations.

In what sense should this approximation be good? Several aspects play a role:

- In many applications, the most important requirement is that the time response $y(t)$ to some given input signal $u(t)$ is approximated by the reduced model.
- Sometimes, not only the output signal $y(t)$ but also the state trajectory $\mathbf{x}(t)$ is to be approximated by the reduced order model.
- As an alternative to approximation in time-domain, we may require that the bode-plot of the original model is well approximated by the reduced model.

- Sometimes, we may require the state variables of the reduced model to have some physical meaning, just like the state variables of the original model.

If we focus on the first aspect for a moment (namely the input-output approximation) then the question is, which "parts" of (1) and, (2) can be neglected without changing the transfer function $g(s)$ significantly. Kalman's controllability and observability concepts give an answer: *uncontrollable and unobservable parts* of the system can be removed without affecting the transfer function. These parts are most easily identified from a diagonal representation[2] of the model (1), (2):

$$\dot{z} = \boldsymbol{T}^{-1}\boldsymbol{A}\boldsymbol{T}z + \boldsymbol{T}^{-1}\boldsymbol{b}u = \begin{bmatrix} \lambda_1 & & \\ & \ddots & \\ & & \lambda_n \end{bmatrix} z + \begin{bmatrix} \bar{b}_1 \\ \vdots \\ \bar{b}_n \end{bmatrix} u, \quad (4)$$

$$y = \mathbf{c}^T \mathbf{T}\mathbf{z} = [\bar{c}_1 \ \ldots \ \bar{c}_n]\mathbf{z}. \quad (5)$$

Figure 3 illustrates, that in this representation the system consists of n independent transfer paths. Obviously, from this model, a state-variable z_i can be removed without affecting the input-output-relation, if either $\bar{b}_i$ is zero (then z_i is said to be uncontrollable), or $\bar{c}_i$ is zero (then z_i is said to be unobservable), or both. By removing unobservable and uncontrollable state-variables, the state-space model becomes a so-called *minimal realization* of the system.

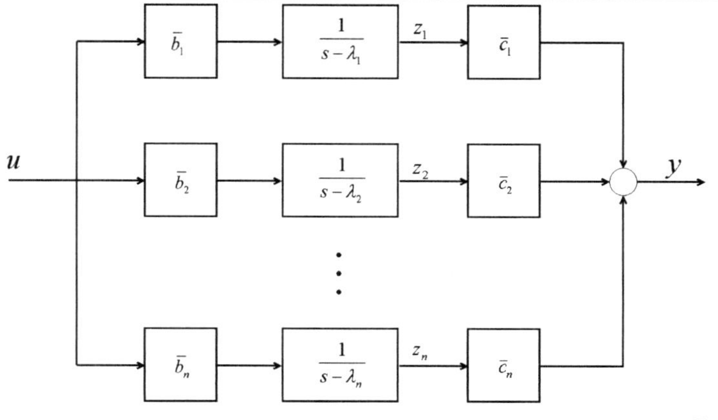

Figure 3 Block diagram of model (4), (5). The ith path can be removed, if either $\bar{b}_i = 0$, or $\bar{c}_i = 0$, or both.

Now, it is only a small step to model order reduction: **The key idea of model order reduction is to not only remove uncontrollable/unobservable parts but to also remove *weakly* controllable/observable parts**. Starting from (4) and (5) this means removing a state-

[2] Equations (4),(5) result from (1),(2) by applying the state transformation **x=Tz** where **T** is the matrix of eigenvectors of **A**. The eigenvalues of **A** are denoted by $\lambda_1,\ldots,\lambda_n$.

variable z_i if $\bar{b}_i$ is small or $\bar{c}_i$ is small or both. These simple steps in fact were the basis of the first reduction methods in state-space, as introduced in the 60s [5]. The more famous method of *balancing and truncation* [14] works in a similar way by using a so-called balanced realization instead of the diagonal representation (4), (5).

As a result, if we assume that the first q state-variables in (4) are to be preserved and the remaining ones are to be removed (this sequence can be achieved by renumbering and rearranging the state variables), the *reduced order model* is

$$\dot{\mathbf{x}}_r = \underbrace{\mathbf{T}_{upper}^{-1}\mathbf{A}\mathbf{T}_{left}}_{\mathbf{A}_r}\mathbf{x}_r + \underbrace{\mathbf{T}_{upper}^{-1}\mathbf{b}}_{\mathbf{b}_r}u, \qquad y = \underbrace{\mathbf{c}^T\mathbf{T}_{left}}_{\mathbf{c}_r^T}\mathbf{x}_r, \tag{6), (7}$$

where $\mathbf{T}_{upper}^{-1}$ comprises the q upper rows of $\mathbf{T}^{-1}$ and $\mathbf{T}_{left}$ comprises the q left columns of $\mathbf{T}$, and $\mathbf{x}_r$ is the state vector of the reduced model. On the right hand side of (6), we now consider the term $\mathbf{T}_{upper}^{-1}\mathbf{A}\mathbf{T}_{left}\mathbf{x}_r$: the vector $\mathbf{x}_r$ of dimension q is first mapped by $\mathbf{T}_{left}$ into a space of dimension n, in which $\mathbf{T}_{left}\mathbf{x}_r$ approximates the original state vector $\mathbf{x}$. Multiplication by $\mathbf{A}$ approximates $\mathbf{Ax}$ by $\mathbf{A}\mathbf{T}_{left}\mathbf{x}_r$. In the last step, the result is mapped back into a space of dimension q, with the result $\mathbf{T}_{upper}^{-1}\mathbf{A}\mathbf{T}_{left}\mathbf{x}_r$. Thereby, the original state-space of dimension n is projected[3] onto the space of the reduced model with dimension q; We perform an *order reduction by projection* [22]. The advantage of the projection result (6), (7) compared to removing rows and columns from (4), (5) is that $\mathbf{T}$ and $\mathbf{T}^{-1}$ need not to be calculated, but only $2q$ columns/rows of them.

More generally, instead of using matrices $\mathbf{T}_{left}$ and $\mathbf{T}_{upper}^{-1}$ (which are biorthogonal, i.e. $\mathbf{T}_{upper}^{-1}\mathbf{T}_{left} = \mathbf{I}$), other (n,q)-matrices **V**, **W** can be used for projection. Furthermore, a more general model of type

$$\mathbf{E}\dot{\mathbf{x}} = \mathbf{A}\mathbf{x} + \mathbf{b}u, \tag{8}$$

$$y = \mathbf{c}^T\mathbf{x}, \tag{9}$$

can be considered instead of (1) and (2). Then, applying the projection matrices **V** and **W** to this model leads to the *general reduced order model by projection:*

$$\underbrace{\mathbf{W}^T\mathbf{E}\mathbf{V}}_{\mathbf{E}_r}\dot{\mathbf{x}}_r = \underbrace{\mathbf{W}^T\mathbf{A}\mathbf{V}}_{\mathbf{A}_r}\mathbf{x}_r + \underbrace{\mathbf{W}^T\mathbf{b}}_{\mathbf{b}_r}u, \tag{10}$$

$$y = \underbrace{\mathbf{c}^T\mathbf{V}}_{\mathbf{c}_r^T}\mathbf{x}_r. \tag{11}$$

Besides delivering these two projection matrices **V** and **W**, an order reduction method should have the following properties:

- It should give some hint on suitable choices of q, the order of the reduced model.
- The reduction should be *invariant to different representations* of the original model (i.e. non-singular state transformations to (1), (2) or multiplication of the model with constant matrices should not influence the transfer function of the reduced model).

---

[3] Mathematically more precisely: $\mathbf{P} = \mathbf{T}_{left}(\mathbf{T}_{upper}^{-1}\mathbf{T}_{left})^{-1}\mathbf{T}_{upper}^{-1}$ is a projection, since $\mathbf{P}^2 = \mathbf{P}$.

- Stable original models should lead to *stable reduced order models*.
- It should be clear in what sense the reduced order model is a "good" or *optimal* approximation. *Error bounds* can help to judge suitability of reduced models.
- Numerical computations should be *simple and numerically robust*. The method should be *automatic*, in the sense that there are not many design parameters to play with.
- The method should be able to handle *general models* like $\mathbf{E}\dot{\mathbf{x}} = \mathbf{A}\mathbf{x} + \mathbf{b}u$ or $\mathbf{M}\ddot{\mathbf{x}} + \mathbf{D}\dot{\mathbf{x}} + \mathbf{K}\mathbf{x} = \mathbf{b}u$.

The *Balancing and Truncation* method [14] fulfils these requirements with some restrictions regarding the last two bullets. The Krylov subspace method to be presented in the next section fulfils the last two bullets but not all of the previous ones (see [16] for invariance properties and see [8] for stability/passivity issues). Ongoing research may improve the Krylov subspace methods in the future and may even deliver combinations of Krylov subspace methods and Balancing/Truncation techniques [1].

In the next section, we introduce the Krylov subspace based reduction and its basic properties. In sections 3 and 4, we present the two most popular algorithms applied in numerical computations; section 5 discusses an interesting extension. Section 6 presents a simple stopping criterion for the iterative algorithms, fixing the order q of the reduced model. Some numerical aspects and an application example can be found in sections 7 and 8. Section 9 gives a short outlook onto future work in linear and nonlinear model reduction.

2 Krylov subspaces in order reduction

The *Krylov subspace* is defined

$$K_q(\widetilde{\mathbf{A}}, \widetilde{\mathbf{b}}) = span\{\widetilde{\mathbf{b}}, \widetilde{\mathbf{A}}\widetilde{\mathbf{b}}, \ldots, \widetilde{\mathbf{A}}^{q-1}\widetilde{\mathbf{b}}\}, \tag{12}$$

where $\widetilde{\mathbf{A}}$ is a constant (n,n)-matrix, $\widetilde{\mathbf{b}}$ is a constant $(n,1)$-vector (the so-called *starting vector*) and q is some given positive integer. The vectors $\widetilde{\mathbf{b}}, \widetilde{\mathbf{A}}\widetilde{\mathbf{b}}, \widetilde{\mathbf{A}}^2\widetilde{\mathbf{b}}, \ldots$ constructing the subspace are called *basic vectors*. With this definition, the two projection matrices, determining the reduced order model (10), (11), are chosen as follows:

$\mathbf{V}$ is any basis of the Krylov subspace $K_{q_1}(\mathbf{A}^{-1}\mathbf{E}, \mathbf{A}^{-1}\mathbf{b}) = span\{\mathbf{A}^{-1}\mathbf{b}, \ldots, (\mathbf{A}^{-1}\mathbf{E})^{q_1-1}\mathbf{A}^{-1}\mathbf{b}\}$,
$\mathbf{W}$ is any basis of the Krylov subspace $K_{q_2}(\mathbf{A}^{-T}\mathbf{E}^{\mathbf{T}}, \mathbf{A}^{-T}\mathbf{c}) = span\{\mathbf{A}^{-T}\mathbf{c}, \ldots, (\mathbf{A}^{-T}\mathbf{E}^{\mathbf{T}})^{q_2-1}\mathbf{A}^{-T}\mathbf{c}\}$,

with **A, E, b, c** from (8), (9), with $\mathbf{A}^{-\mathbf{T}} = (\mathbf{A}^{-1})^T$, and with q_1, q_2 such that **V** and **W** both are rank q [4]. With this choice (and provided that the resulting $\mathbf{A}_r$ is non-singular) the reduction method is called a *two-sided* method, since input as well as output relations of the system are involved. If only one of the two matrices **V, W** is chosen as specified and the other one is chosen arbitrarily but such that $\mathbf{A}_r$ is non-singular, then the method is called a *one-sided* method.

The suggested choices of **V** and **W** look very simple. However, the numerical calculation of the matrix-vector-products involved in the Krylov subspaces turns out to tend to be unstable. Therefore, algorithms have been developed to reliably calculate **V** and **W** (see next sections).

---

[4] If some desired value of q can not be achieved, the reason is that the controllability or the observability subspace has a dimension smaller than q. Then, q must be reduced.

An important question now is: In what sense is the above choice of $\mathbf{V}$ and $\mathbf{W}$ "good"? In fact, certain parameters of the input-output relation of the original and of the reduced model equal each others, the so-called *moments*. The moments m_i are defined as the (negative) coefficients of the taylor-series of the transfer function (3) around $s_0 = 0$,

$$g(s) = \mathbf{c}^{\mathrm{T}}(s\mathbf{I}-\mathbf{A})^{-1}\mathbf{b} = -\underbrace{\mathbf{c}^T\mathbf{A}^{-1}\mathbf{b}}_{m_0} - \underbrace{\mathbf{c}^T(\mathbf{A}^{-1}\mathbf{E})\mathbf{A}^{-1}\mathbf{b}}_{m_1}s - \ldots - \underbrace{\mathbf{c}^T(\mathbf{A}^{-1}\mathbf{E})^i\mathbf{A}^{-1}\mathbf{b}}_{m_i}s^i - \ldots \, . \quad (13)$$

In two-sided methods, the first $2q$ moments of the original and the reduced model match, in one-sided methods q moments match. In order to give an idea of how this can be proved, we consider the first moment, m_{r0}, of the reduced model (10), (11) and show that it equals m_0:

$$m_{r0} = \mathbf{c}_r^T\mathbf{A}_r^{-1}\mathbf{b}_r = \mathbf{c}^T\mathbf{V}(\mathbf{W}^T\mathbf{A}\mathbf{V})^{-1}\mathbf{W}^T\mathbf{b} = \mathbf{c}^T\mathbf{V}(\mathbf{W}^T\mathbf{A}\mathbf{V})^{-1}\mathbf{W}^T\mathbf{A}\mathbf{V}\mathbf{r}_0 = \mathbf{c}^T\mathbf{V}\mathbf{r}_0 = \mathbf{c}^T\mathbf{A}^{-1}\mathbf{b} = m_0 \, . \quad (14)$$

The essential step in this proof is that $\mathbf{A}^{-1}\mathbf{b}$ belongs to the Krylov subspace $K_{q_1}(\mathbf{A}^{-1}\mathbf{E},\mathbf{A}^{-1}\mathbf{b})$ and therefore can be expressed as $\mathbf{V}\mathbf{r}_0$; thereby, $\mathbf{b} = \mathbf{A}\mathbf{V}\mathbf{r}_0$ and $\mathbf{V}\mathbf{r}_0 = \mathbf{A}^{-1}\mathbf{b}$. For the higher moments, the steps of proof are similar [16], using the facts that $(\mathbf{A}^{-1}\mathbf{E})^i\mathbf{A}\mathbf{b} = \mathbf{V}\mathbf{r}_i$ and $(\mathbf{A}^{-T}\mathbf{E}^T)^i\mathbf{A}^{-T}\mathbf{c} = \mathbf{W}\mathbf{q}_i$.

As a result, we can state that the reduced model approximates the original one in the sense that the first q or $2q$ coefficients of the taylor series of $g(s)$ around zero match. Some additional properties of the reduction are:

- Instead of $s_0 = 0$, the taylor series can be considered at other points $s_0 \neq 0$. Then, with slight modifications of the Krylov subspaces, the moments can be matched there. It is even possible to match moments for different values s_0 at the same time, and even for $s_0 \to \infty$ (the moments are called *Markov parameters*, then), see [10,16,22].
- In one-sided methods, typically $\mathbf{V}$ is chosen as a basis of $K_{q_1}(\mathbf{A}^{-1}\mathbf{E},\mathbf{A}^{-1}\mathbf{b})$. Then, it can be shown (by a similar line as (14)) that for *any* output y, q moments match. Furthermore, typically $\mathbf{W} = \mathbf{V}$ is chosen. In this case, for certain original models with passivity-related properties, it is possible to proove that the reduced model will be passive [8]. Another advantage of one-sided methods with $\mathbf{W} = \mathbf{V}$ is that they often deliver better approximations of the original state trajectory by $\hat{\mathbf{x}}(t) = \mathbf{V}\mathbf{x}_r(t)$, than methods with $\mathbf{W} \neq \mathbf{V}$ do.
- Two-sided methods have the advantage of more matching moments, leading to better approximations of the output y in many applications. Furthermore, the input-output behaviour of a reduced model found by a two-sided method is independent of the model representation we start from. This is not true for one-sided methods, where non-singular state-transformations or matrix multiplications applied to (8), (9) influence the reduction result.
- In general, there is no guarantee for stability of the reduced model. When reducing the destillation column from figure 1 we experienced that one-sided methods did not deliver stable models at all, while some two-sided approaches worked. When reducing the micro-fluidic system from figure 2, all one-sided methods delivered stable models (because the preconditions for passivity of the reduced model [8] are fulfilled in this system), while two-sided methods for some q delivered unstable reduced models.

3 Arnoldi algorithm

In order to avoid numerical problems when building up the Krylov subspace, it is an advantage to construct an *orthogonal* basis for a given subspace. The classical Arnoldi procedure [2] finds a set of orthonormal vectors as a basis for a given Krylov subspace. In the following, we briefly describe the Arnoldi algorithm.

Consider the Krylov subspace $K_q(\mathbf{A}_1, \mathbf{b}_1)$; the Arnoldi algorithm finds a set of vectors with norm one that are orthogonal to each others,

$$\mathbf{V}^T\mathbf{V} = \mathbf{I}, \tag{15}$$

where the columns of the matrix $\mathbf{V}$ are the basis for the given Krylov subspace. Furthermore,

$$\mathbf{V}^T\mathbf{A}_1\mathbf{V} \quad = \quad \mathbf{H},$$

where $\mathbf{H} \in \mathbb{R}^{q\times q}$ is Hessenberg[5]. In the following algorithm, in each step one more vector, orthogonal to all other previous vectors, is constructed and normalized. It is assumed that in each step in fact an orthogonal vector exists (should this not be true, then the algorithm can not be continued and q must be reduced).

Algorithm 1 *Arnoldi algorithm (using modified Gram-Schmidt Orthogonalization) [8]:*

0. ***Start:*** *Set*

$$\mathbf{v}_1 = \frac{\mathbf{b}_1}{\sqrt{\mathbf{b}_1^T\mathbf{b}_1}},$$

where $\mathbf{b}_1$ is the starting vector, then set $i = 2$.

1. ***Calculating the next vector:*** *The next vector is*

$$\hat{\mathbf{v}}_i = \mathbf{A}_1\mathbf{v}_{i-1}.$$

2. ***Orthogonalization:*** *For j=1 to i-1 do:*

$$h = \hat{\mathbf{v}}_i^T\mathbf{v}_j \quad , \quad \hat{\mathbf{v}}_i = \hat{\mathbf{v}}_i - h\mathbf{v}_j.$$

3. ***Normalization:*** *If vector $\hat{\mathbf{v}}_i$ is zero, break the loop. Else, the i-th column of matrix $\mathbf{V}$ is*

$$\mathbf{v}_i \quad = \quad \frac{\hat{\mathbf{v}}_i}{\sqrt{\hat{\mathbf{v}}_i^T\hat{\mathbf{v}}_i}}.$$

4. ***Check for break down:*** *Stop if $i = q$, otherwise increase i and go to step 1.*

By knowing that in each step of the Arnoldi algorithm, the i-th column of $\mathbf{V}$ is a linear combination of the first i basic vectors, it is not difficult to show that the vectors $\{\mathbf{v}_1, \cdots, \mathbf{v}_i\}$ span the same space as the first i basic vectors of the given Krylov subspace.

In one-sided Krylov subspace methods, the Arnoldi algorithm is the most common way to find a basis for the input or output Krylov subspace and compute the matrix $\mathbf{V}$ or $\mathbf{W}$ respectively. For instance, by using the input Krylov subspace $K_q(\mathbf{A}^{-1}\mathbf{E}, \mathbf{A}^{-1}\mathbf{b})$ and applying the algorithm 1, the matrix $\mathbf{V}$ can be found and matrix $\mathbf{W}$ can be chosen arbitrarily but such that $\mathbf{A}_r$ is nonsingular (a "good" choice often is $\mathbf{W} = \mathbf{V}$). The reduced order model then is given by (10) and (11), with q moments matching. Note however, that the transfer function of the reduced model depends on the particular choice of $\mathbf{W}$, [16].

[5]It means that $h_{ij} = 0$ for $i > j + 1$ for all $1 \leq i, j \leq q$

4 Modified Lanczos algorithm

A well known algorithm in two-sided Krylov methods is Lanczos algorithm. The classical Lanczos algorithm [12] and its improved version in [8] generate two sequences of basis vectors which span the Krylov subspaces $K_q(\mathbf{A}_1, \mathbf{b}_1)$ and $K_q(\mathbf{A}_1^T, \mathbf{c}_1)$ such that,

$$\mathbf{W}^T\mathbf{V} = \mathbf{I}, \tag{16}$$

where the columns of the matrix $\mathbf{V}$ and $\mathbf{W}$ are the bases for $K_q(\mathbf{A}_1, \mathbf{b}_1)$ and $K_q(\mathbf{A}_1^T, \mathbf{c}_1)$, respectively. This algorithm is limited to the special case that the matrix $\mathbf{A}$ is equal to identity. Otherwise, it is necessary to multiply the state equation in system (8,9) by $\mathbf{A}^{-1}$ and then apply the Lanczos method for this new system. The Lanczos algorithm is numerically unstable and after some iterations biorthogonality of the vectors is lost. To avoid this problem, reorthogonalization must be used [4]. In this paper, we present a new generalized numerically stable algorithm (see also [17]) using modified Gram-Schmidt (MGS) for the general representation of systems. So, the algorithm can be directly applied to the system (8) and (9).

Consider the Krylov subspaces, $K_q(\mathbf{A}_1, \mathbf{b}_1)$ and $K_q(\mathbf{A}_2, \mathbf{c}_2^T)$ both with the same rank. The Lanczos algorithm can be expressed as follows:

Algorithm 2 *Lanczos algorithm using MGS:*

0. ***Start:*** *Set*

$$\mathbf{v}_1 = \frac{sign(\mathbf{c}_1^T\mathbf{b}_1)\mathbf{b}_1}{\sqrt{|\mathbf{c}_1^T\mathbf{b}_1|}}, \ \mathbf{w}_1 = \frac{\mathbf{c}_1^T}{\sqrt{|\mathbf{c}_1^T\mathbf{b}_1|}}.$$

where $\mathbf{b}_1$ and $\mathbf{c}_1^T$ are the starting vectors.

1. ***Calculating the next vector:*** *The next vectors are computed as follows,*

$$\hat{\mathbf{v}}_i = \mathbf{A}_1\mathbf{v}_{i-1} \quad , \quad \hat{\mathbf{w}}_i = \mathbf{A}_2\mathbf{w}_{i-1}.$$

2. ***Orthogonalization:*** *By using the vectors in the last step and MGS, the new columns before normalization can be found as follows, for j=1 to i-1 do:*

$$h_1 = \hat{\mathbf{v}}_i^T\mathbf{w}_j \quad , \quad \hat{\mathbf{v}}_i = \hat{\mathbf{v}}_i - h_1\mathbf{v}_j,$$
$$h_2 = \hat{\mathbf{w}}_i^T\mathbf{v}_j \quad , \quad \hat{\mathbf{w}}_i = \hat{\mathbf{w}}_i - h_2\mathbf{w}_j.$$

3. ***Normalization:*** *The i-th columns of matrices $\mathbf{V}$ and $\mathbf{W}$ are*

$$\mathbf{v}_i = \frac{sign(\hat{\mathbf{v}}_i^T\hat{\mathbf{w}}_i)\hat{\mathbf{v}}_i}{\sqrt{|\hat{\mathbf{v}}_i^T\hat{\mathbf{w}}_i|}}, \ \mathbf{w}_i = \frac{\hat{\mathbf{w}}_i}{\sqrt{|\hat{\mathbf{v}}_i^T\hat{\mathbf{w}}_i|}}.$$

4. ***Check for break down:*** *If $i = q$ stop otherwise go to step 1 and continue.*

The property of the algorithm 2 is that,

$$\mathbf{W}^T\mathbf{A}_1\mathbf{V} = \mathbf{H}, \tag{17}$$
$$\mathbf{V}^T\mathbf{A}_2\mathbf{W} = \mathbf{T}, \tag{18}$$

where $\mathbf{H}, \mathbf{T} \in \mathbb{R}^{q\times q}$ are Hessenberg matrices. In order reduction, when one of the matrices $\mathbf{A}$ or $\mathbf{E}$ is an identity matrix, $\mathbf{A}_1 = \mathbf{A}_2^T$, and then matrices $\mathbf{H}$ and $\mathbf{T}$ will be transposed of each other. In this case, $h_{ij} = 0$ for $i > j + m$ and $j > i + p$ for all $1 \le i, j \le q$. By this assumption in SISO systems the matrices $\mathbf{H}$ and $\mathbf{T}$ are tridiagonal matrices, which is the famous result in the classical Lanczos algorithm.

5 Two-sided Arnoldi algorithm

In this section we introduce another method called two-sided Arnoldi to find the bases necessary for projection and calculating the reduced order model. This new method in comparison to Lanczos is *numerically stable* and *easier to implement* while leading to the same result as Lanczos [16, 10, 7]. Our algorithm comprises the following steps:

Algorithm 3 *Two-sided Arnoldi algorithm:*

0. *Choose the appropriate input and output Krylov subspaces for the given system, i.e.* $K_{q_1}(\mathbf{A}^{-1}\mathbf{E}, \mathbf{A}^{-1}\mathbf{b})$ *and* $K_{q_2}(\mathbf{A}^{-T}\mathbf{E}^T, \mathbf{A}^{-T}\mathbf{c})$.
1. *Apply Arnoldi algorithm 1 to the input Krylov subspace to find the matrix* $\mathbf{V}$.
2. *Apply Arnoldi algorithm 1 to the output Krylov subspace to find the matrix* $\mathbf{W}$.
3. *Find the reduced order model from equation (10) and(11).*

For comparison: in the Lanczos algorithm, the vectors of each basis are orthogonal to the vectors in the other basis except for one of them,

$$\mathbf{V}^T\mathbf{W} = \mathbf{I},$$

whereas in the two-sided Arnoldi, each basis is orthonormal,

$$\mathbf{V}^T\mathbf{V} = \mathbf{I} \text{ and } \mathbf{W}^T\mathbf{W} = \mathbf{I}.$$

6 Stopping criteria

One of the important parts in order reduction is to find a suitable order, q, for the reduced model which leads to a good approximation. Roughly speaking, the question is: in iterative order reduction when can we stop the iteration loop? To answer this question one way is to find a reduced order model and then compare it to the original system. This comparison can be done for instance by computing the norm of the error between two transfer functions. By knowing that calculating the 2-norm or infinity-norm of a large scale model using the current algorithms is very time consuming or even impossible, finding an easy to calculate measure that can be examined in each iteration seems to be essential. In iterative methods, it is desired to have a measure that can be tested in each iteration, without breaking the loop. If the errors are acceptable we can stop, otherwise higher order is necessary.

As mentioned before, in a Krylov subspace, if the i-th basic vector is a linear combination of the first $i-1$-st basic vectors, then all the next basic vectors are also linear combinations of the first $i-1$-st basic vectors. In that case Arnoldi algorithm 1 finds the vector $\hat{\mathbf{v}}_i = 0$ and a basis for the given subspace is found.

Now, consider the original model is not controllable. If the Arnoldi algorithm is applied to such a model to find a basis for input Krylov subspace, by continuing the iterations to find zero vector in step 2 of algorithm 1, it is possible to delete the non-controllable modes in the reduced order model. The same result to delete the non-observable modes can be obtained by applying the Arnoldi algorithm to the output Krylov subspace. In two-sided Arnoldi, if one of the vectors before normalization, in the basis of input or output Krylov subspace becomes zero then we can stop and a minimal realization can be found.

The main goal of order reduction can be deleting the *almost* non-controllable and non-observable parts. In the Krylov subspace, if the i-th basic vector is *almost* linear combination of the first $i-1$-st basic vectors (i.e. there exists a linear combination of them that is very small) then, we can stop the iterations. In that case Arnoldi algorithm 1 finds the vector $\hat{\mathbf{v}}_i \approx 0$.

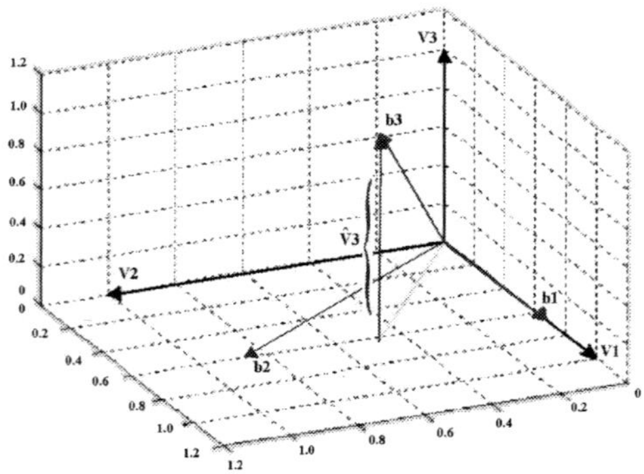

Figure 4: Arnoldi algorithm in 3 dimensional view.

Figure 4 is a 3 dimensional aspect of Arnoldi algorithm until finding the 3rd vector. Vector $\mathbf{b}_1$ is the starting vector but the algorithm chooses the normalized form of it as the first element of the basis. $\mathbf{b}_2 = \mathbf{A}_1\mathbf{v}_1$ and the algorithm chooses the vector $\mathbf{v}_2$. In the next step, if the vector $\mathbf{b}_3$ lies on the plane of vectors $\mathbf{v}_1$ and $\mathbf{v}_2$, then we can stop but, if this vector has a very small angle with that plane (sine of this angle can be calculated as $\frac{norm(\hat{\mathbf{v}}_3)}{norm(\mathbf{b}_3)}$) or the vector $\hat{\mathbf{v}}_3$ is very small then we can consider that the vectors $\mathbf{v}_1$, $\mathbf{v}_2$ and $\mathbf{b}_3$ are almost linearly dependent.

Therefore, two measures can be used as stopping criteria:

1. Sine of the angle between the vector $\mathbf{A}_1\mathbf{v}_{i-1}$ and the plane (it can be a space or hyperspace) constructed by all previous vectors, for input Krylov subspace it is $d_1 = \frac{norm(\hat{\mathbf{v}}_i)}{norm(\mathbf{A}^{-1}\mathbf{E}\mathbf{v}_{i-1})}$.
2. The norm of the new vector before normalization, $d_2 = norm(\hat{\mathbf{v}}_i)$.

In two-sided Arnoldi, multiplication of the two measures for input and output Krylov subspaces can be considered, for example $norm(\hat{\mathbf{v}}_i) \cdot norm(\hat{\mathbf{w}}_i)$. When this measure is small or falls down and remains constant after some iterations, then we can stop and calculate the reduced order model. However, there is no guaranty that for all cases the method works well or the measure falls down after some iterations.

7 Inverse of a large matrix

In order reduction of the system (8) and (9) based on moment matching the matrix $\mathbf{A}^{-1}\mathbf{E}$ and $\mathbf{A}^{-T}\mathbf{E}^T$ are very important. So, the inverse of the large matrix $\mathbf{A}$ seems to be necessary. Calculating this inverse and then use it in the algorithm 1 is not recommended for numerical reasons. In order to find the vector $\mathbf{x} = \mathbf{A}^{-1}\mathbf{E}\mathbf{v}_{i-1}$, it is better to solve the linear equation $\mathbf{A}\mathbf{x} = \mathbf{E}\mathbf{v}_{i-1}$ for $\mathbf{x}$ in each iteration to avoid numerical errors and to save time. (In this way, a total of q sets of linear equations are solved, whereas the calculation of matrix $\mathbf{A}^{-1}$ would require n sets of such equations.)

There exist many methods that can solve a linear equation and find an exact or approximate solution. One of the best method to find an exact solution is using LU-factorization [9] and then solve two triangular linear equations by Gaussian elimination. Using this method in *each* iteration leads to a slow algorithm while the most time consuming part is finding the LU factorization of the large matrix. The remedy is finding the LU factorization of the matrix $\mathbf{A}$ *at the beginning* and then solve only triangular linear equations in each iteration [6]. In this case, time is saved and the result is obtained very fast.

8 Numerical example

Two methods have been applied to reduce the order of several systems:

1. One-sided Krylov subspace method using Arnoldi algorithm applied to input Krylov subspace and choosing $\mathbf{W} = \mathbf{V}$. In this case there is no need to define an output.

2. Two-sided Krylov subspace using the two-sided Arnoldi and Lanczos algorithms (they lead to the same results) to match more moments and improve the results.

The first example is a class of high energy Micro-Electro-Mechanical System (MEMS) actuators which integrates solid fuel with three silicon micromachined wafers [3] as shown in figure 2. A two-dimensional axi-symmetric model is used, which after finite element based spatial discretization of the governing equations results in a linear system of $n = 1071$ ordinary differential equations. As expected, the two-sided method performs better in approximating the output (node number 1) and in this particular example even performs better when comparing all the state variables instead of the output only [20]. No steady state error occurs; this is due to matching of the first moment, responsible for the "DC gain" of the reduced model, [16]. It is to be mentioned that for the system considered here, the one-sided methods (with $\mathbf{W} = \mathbf{V}$) always lead to stable reduced models whereas the two-sided methods can produce unstable models. More details can be found in [20].

The second example is the beam model which is SISO with 348 states [6][7]. The model is obtained by spatial discretization of an appropriate partial differential equation. The input represents the force applied to the structure at the free end, and the output is the resulting displacement. The multiplication two measures d_2 from two Arnoldi algorithms in two-sided Arnoldi is used as stopping criterion, as shown in figure 5. Based on the stopping criterion, order 6 is suitable for the reduced system. By this reduced model, the step response of the original system can be approximated very good, as shown in figure 6.

Finally, different reduction methods have been applied to the distillation column (figure 1, a model with order 400). It turned out that balancing and truncation can be applied and reduced to order 30 with excellent results. Alternatively, Krylov subspace methods have been applied: one-sided input subspace approaches led to unstable reduced models while the two-sided approach of section 5 led to good results with a reduced order of 46, using a selection procedure in the MIMO implementation of the Arnoldi algorithm [18].

9 Conclusion and Outlook

In the previous sections, we have introduced:

(i) Some basic concepts of linear model order reduction (for balancing and truncation see [14] and for modal reduction see [5]).

(ii) The essential relations in Krylov subspace based projections.

(iii) The most commonly used algorithms for the numerical implementation, together with: (a) a stopping criterion and (b) some numerical issues.

---

[6] In MATLAB instead of the command `A\V(:,i-1)`, the command `[L,U]=LU(A)` must be added to the beginning of the algorithm and in each iteration the command `U\(L\V(:,i-1))` can be used to solve triangular linear equations.

[7] The model is available online at http://www.win.tue.nl/niconet/NIC2/benchmodred.html.

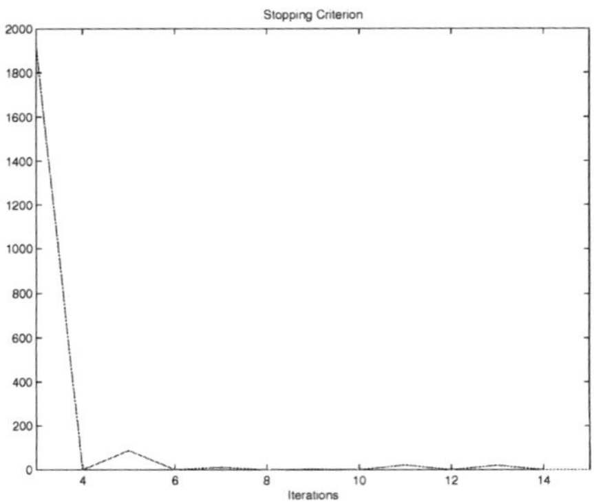

Figure 5: Stopping criterion for beam model using two-sided Arnoldi.

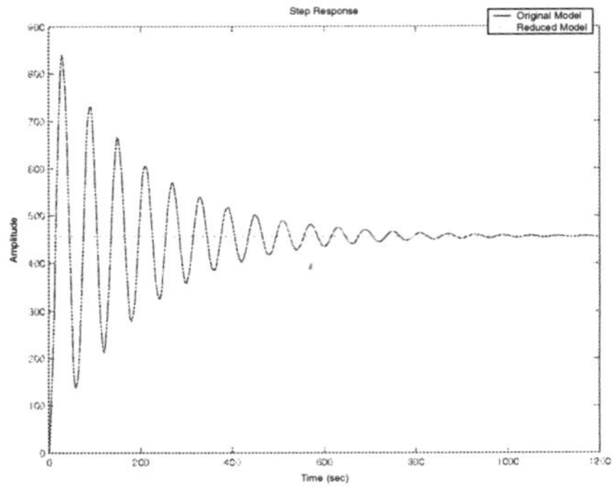

Figure 6: Step response of the original beam model and its 6 order approximation using two-sided Arnoldi.

Due to their simplicity, Krylov subspace based reduction schemes can be applied to very large order systems with up to ten or even hundred thousand state variables. Their disadvantages (compared to the very successful method of Balancing and Truncation) are the facts that there is no guarantee for stability of the reduced model and that there are no error bounds known today. Therefore, recent research goes towards approximating the Balancing and Truncation results using Krylov subspace methods [1].

Another challenging field of research is the reduction of nonlinear system models. One approach is to apply the projections as introduced in section 1 to nonlinear models resulting in reduced models of the form, where matrix $\mathbf{V}$ is gained from an analysis of simulation data of the original model or from reducing a linearized model. Such approaches are referred to as Proper Orthogonal Decomposition (POD) methods [11]. Other successful approaches are based on isolating the nonlinearities and reducing the linear system parts around them [19, 13]. These approaches and POD approaches, both, are essentially based on linear considerations, without guaranteed error bounds for the nonlinear model, but applicable to medium or even large-scale models [19]. The nonlinear approach of balancing and truncation by [21, 15], on the other hand, is limited to small systems for computational reasons.

References

[1] Antoulas, A. C. and Sorensen, D. Approximation of Large-Scale Dynamical Systems: an Overview. *Int. J. Applied Math. Computer Science*, 11(5):1093–1121, 2001.

[2] Arnoldi, W. E. The Principle of Minimized Iterations in Solution of the Matrix Eigenvalue Problem. *Quarrt. Appl. Math.*, 9:17–29, 1951.

[3] Bechtold, T., Rudnyi, B., and Korvink, J. Automatic Order reduction of Thermo-Electric Models for MEMS: Arnoldi versus Guyan. Proc. 4th ASDAM, 2001.

[4] Boley, D. L. Krylov space methods on state-space control models. *Circuits Syst. Signal Process*, 13:733–758, 1994.

[5] Bonvin, D. and Mellichamps, D. A. A Unifind Derivation and Critical Review of Modal Approaches to Model Reduction. *Internation Journal of Control*, 35:829–848, 1982.

[6] Chahlaoui, Y. and Van Dooren, P. A collection of benchmark examples for model reduction of linear time invariant dynamical systems. SLICOT Working Note 2002-2, 2002.

[7] Cullum, J. and Zhang, T. Two-sided Arnoldi and Nonsymmetric Lanczos Algorithms. *SIAM Journal on Matrix Analysis and Applications*, 24(2):303–319, 2002.

[8] Freund, R. W. Passive Reduced-Order Modeling via Krylov Subspace Methods. *Numerical Analysis Manuscript*, (00-3-02), March 2000.

[9] Golub, G. H. and Van Loan, C. F. *Matrix Computations*. The Johns Hopkins University Press, 3rd edition, 1996.

[10] Grimme, E. J. *Krylov Projection Methods for Model Reduction*. PhD thesis, Dep. of Electrical Engineering, University of Illinois at Urbana Champaign, 1997.

[11] Lall, S, Marsden, J. E., and Glavaski, S. Empirical model reduction of controlled nonlinear systems. Technical Reports CIT/CDS 98-008, California Institute of Technology, 1998.

[12] Lanczos, C. An Iteration Method for the Solution of the Eigenvalue Problem of Linear Differential and Integral Operators. *J. Res. Nat. Bureau Stan.*, 45:255–282, 1950.

[13] Lohmann, B. Order Reduction and Determination of Dominant State Variables of Nonlinear Systems. *Mathematical Modelling of Systems*, 1(2):77–90, 1995.

[14] Moore, B.C. Principal Component Analysis in Linear Systems: Controllability, Observability and Model Reduction. *IEEE Trans. on Automatic Control*, AC-26:17–31, 1981.

[15] Newman, J. A. and Krishnaprasad, P. S. Computing Balanced Realizations for Nonlinear Systems. Tech. Rep. CDCSS TR 2000-4, Inst. Systems Research, Uni. Maryland, 2000.

[16] Salimbahrami, B. and Lohmann, B. Krylov Subspace Methods in Linear Model Order Reduction: Introduction and Invariance Properties. Scientific report, 2002. Available at http://www.iat.uni-bremen.de/mitarbeiter/salimbahrami/Invariance_properties.pdf.

[17] Salimbahrami, B. and Lohmann, B. Modified Lanczos Algorithm in Model Order Reduction of MIMO Linear Systems. Sci. rep., Institute of Automation, University of Bremen, 2002. Available at http://www.iat.uni-bremen.de/mitarbeiter/salimbahrami/mlanczos.pdf.

[18] Salimbahrami, B. and Lohmann, B. Two-sided Arnoldi Method for Model Order Reduction Of High Order MIMO Systems. Sci. rep., 2002. Available at http://www.iat.uni-bremen.de/mitarbeiter/salimbahrami/2arnoldimimo.pdf.

[19] Salimbahrami, B. and Lohmann, B. A Simulation Free Nonlinear Model Order Reduction Approach and Comparison Study. In *Proc. 4th Mathmod*, pages 429–435, Vienna, 2003.

[20] Salimbahrami, B., Lohmann, B., Bechtold, T., and Korvink, J. G. Two-sided Arnoldi Algorithm and Its Application in Order Reduction of MEMS. In *Proc. 4th Mathmod*, pages 1021–1028, Vienna, 2003.

[21] Scherpen, J. M. A. *Balancing for Nonlinear Systems*. Ph.D. Dissertation, University of Twente, The Netherlands, 1994.

[22] Villemagne, C. D. and Skelton, R. E. Model reduction using a projection formulation. *Int. J. Control*, 46:2141–2169, 1987.

Applications of Zubov's method to perturbed systems

Fabian Wirth
Zentrum für Technomathematik, Universität Bremen
Postfach 330440, 28334 Bremen
E-mail: fabian@math.uni-bremen.de

Abstract

We describe Zubov's method for the calculation of domains of attraction and a recent generalization of this method that is applicable for systems subject to perturbations. Some examples are given to illustrate the method.

1 Introduction

In the analysis of nonlinear systems local stability analysis at fixed points is quite often a fairly easy task. However, depending on the problem at hand local considerations may not yield sufficient information to determine whether the system has the global stability properties that are needed. For instance, when a nonlinear system is stabilized at a fixed point using local techniques it may be hard to check if the designed controller drives all initial conditions of interest to the desired equilibrium position.

In many applications involving nonlinear systems it is unclear how to achieve global stabilization so that local design methods are used. Under these circumstances global properties are particularly interesting but also particularly hard to come by. The object of interest in this case is the domain of attraction of an asymptotically stable fixed point x^*, that is, the set of all initial conditions for which the trajectory converges to x^*. For the fundamentals of the corresponding theory we refer to the books [7, 8] and further information is to be obtained in the articles [13, 14].

Unfortunately, the domain of attraction is often difficult to describe and it is therefore of interest to be able to give estimates for it. One way to obtain such estimates relies on the use of local Lyapunov functions. The problem of finding appropriate Lyapunov functions and of determining the sublevel sets of such a Lyapunov function that can be guaranteed to be a subset of the domain of attraction has received widespread attention in recent years. In particular, there are numerous investigations of domains of attractions for polynomial systems, see [6, 10, 12, 13] and references therein. However, a priori, a Lyapunov function only gives a conservative estimate for the domain of attraction in that a subset of the domain is described. It is often not at all evident how good an approximation has been obtained using such a method.

In this note we present a method that was originally developed by Zubov [15] which allows in principle for the exact calculation of the whole domain of attraction and we describe extension to perturbed systems. We begin by summarizing Zubov's result. Assume we are given a system of the form

$$\dot{x} = f(x), \quad x \in \mathbb{R}^n \tag{1}$$

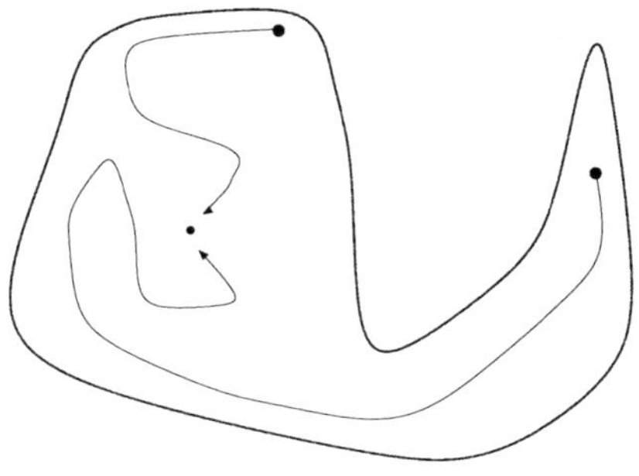

Figure 1: Domain of attraction

with an asymptotically stable fixed point x^*. We assume throughout that $x^* = 0$ and that the vector field f is locally Lipschitz continuous on $\mathbb{R}^n$, so that existence and uniqueness of solutions is guaranteed. By $\varphi(t;x_0)$ we denote the solution at time t of (1) satisfying the initial condition $\varphi(0;x_0) = x_0$. The object of interest is now the domain of attraction (of the asymptotically stable fixed point $x^* = 0$), which is defined as follows

$$\mathcal{D}(0) := \{x \in \mathbb{R}^n \mid \varphi(t;x) \to 0, \text{ for } t \to \infty\} .$$

Figure 1 represent a sketch of a domain of attraction. Unfortunately there are few restrictions as to what a domain of attraction can look like. Basically any open subset of $\mathbb{R}^n$ that is diffeomorphic to $\mathbb{R}^n$ can be a domain of attraction. For details, we refer to [7, 14]. Thus the question of computation of such domains becomes crucial.

One of the milestones in the analysis of domains of attraction is Zubov's theorem [15] which links the domain of attraction of an asymptotically stable fixed point $x^* = 0$ of (1) to the first order partial differential equation

$$Dv(x) \cdot f(x) = -h(x)(1 - v(x))\sqrt{1 + \|f(x)\|^2}, \tag{2}$$

where h is a positive auxiliary function. Namely, under suitable assumptions on h, equation (2) has a unique solution v that vanishes in the origin. Furthermore, the set $v^{-1}([0,1))$ is equal to the domain of attraction $\mathcal{D}(0)$, or equivalently

$$\mathcal{D}(0) = \{x \in \mathbb{R}^n \mid 0 \leq v(x) < 1\}.$$

From (2) we also see that $Dv(x) \cdot f(x) < 0$ if $v(x) < 1$, due to the positivity of h. From which we can conclude that v is a Lyapunov function for system (1) on $\mathcal{D}(0)$. The importance of this result is twofold. On the one hand it is an *existence* result, which states that a Lyapunov function exists that characterizes the whole domain of attraction. And in fact, there exist many of them since we have freedom in the choice of the auxiliary function h which will in general lead to different solutions of (2). On the other hand the equation also opens the door for computational approaches, as in order to find the Lyapunov function, we "only" have to solve equation (2).

These results are presented in several books, see [7] or [8]. For the case of real-analytic systems a constructive procedure is presented in [7] that allows for the approximation of the domain of attraction. This method was extended and simplified in [13], where again a constructive approach for the case of analytic systems is presented.

In this note we present some recent results obtained in [2, 4] which show an extension of Zubov's result for the case of perturbed systems. In the following Section 2 we present the class of systems under consideration, we introduce the concept of robust domains of attraction and give a brief overview over the theoretical results. In Section 3 some examples are presented where for simple systems robust domains of attraction are calculated. We comment on the merits and the limitations of the method in Section 4.

2 Perturbed Systems

We consider perturbed systems of the form

$$\dot{x}(t) = f(x(t), a(t)), \tag{3}$$

where $a : \mathbb{R} \to A \subset \mathbb{R}^m$ is a measurable function taking values in the compact set A. Here $x \in \mathbb{R}^n$ denotes the state of the system and a represents an unknown time-varying perturbation acting on the system and $f : \mathbb{R}^n \times A \to \mathbb{R}^n$. In order to ensure existence and uniqueness of solutions we assume that f in (3) is continuous and locally Lipschitz continuous in x where the Lipschitz constants can be chosen independently of $a \in A$. We are interested in stability properties that are robust with respect to unknown perturbations taking values in A, or in other words in stability properties that hold for all functions in the set

$$\mathcal{A} := \{a : \mathbb{R} \to A \mid a \text{ is measurable}\}.$$

The solution of (3) corresponding to the initial condition $\varphi(0) = x_0$ and to a particular choice for the perturbation function $a \in \mathcal{A}$ is denoted by $\varphi(t; x_0, a)$. The Euclidean norm in $\mathbb{R}^n$ is denoted by $\|\cdot\|$.

As we are interested in domains of attraction we assume that the point $x^* = 0$ has a robust local stability property. To be precise, we need the following assumption.

Assumption 1 *Consider system* (3). *We assume that there is a ball of radius $r > 0$ around $x^* = 0$ and constants $M \geq 1, \beta < 0$ such that for all $a(\cdot) \in \mathcal{A}$ and all x with $\|x\| \leq r$ we have*

$$\|\varphi(t; x, a)\| \leq M e^{\beta t} \|x\|, \quad t \geq 0.$$

It may now seem that there are two very strict assumptions here that make the problem as stated here a not very interesting one. First of all we have implicitly assumed that the fixed point $x^* = 0$ is not perturbed by any of the perturbations, or in other words that

$$f(0, a) = 0, \quad \text{for all } a \in A,$$

a case that is rarely met in applications. Secondly, we have assumed local exponential decay which is maybe a more reasonable assumption but still it gives rise to the question if this assumption is really necessary.

Indeed, the problem can be studied in a more general context and the results that we present here remain essentially the same. We restrict ourselves to the present case for ease of exposition. For the more general case of compact attractors which includes the case that a fixed point is perturbed by the perturbations and for the case of general attraction rates we refer to [2, 3, 5].

We now define the domain of attraction robust with respect to perturbations.

Definition 2 *If the system* (3) *satisfies Assumption 1 we define the* uniform robust domain of attraction *of* $x^* = 0$ *by*

$$\mathcal{D}_0 = \left\{ x_0 \in \mathbb{R}^n \mid \begin{array}{l} \text{there exists a function } \beta \text{ with } \beta(t) \to 0 \text{ as } t \to \infty \\ \text{such that } \|\varphi(t;x_0,a)\| \le \beta(t) \text{ for all } t > 0,\, a \in \mathcal{A} \end{array} \right\}. \quad (4)$$

It may seem surprising that in the definition of the robust domain of attraction we require a uniform attraction rate given by the function β that is an upper bound for the decay for all perturbations. Indeed, it may be shown that this is not at all a severe restriction. It follows from [4, Prop. 2.3 (v)] that if x_0 is a point such that all trajectories $\varphi(t;x_0,a)$ converge to zero, but an upper bound β as in the definition of $\mathcal{D}_0$ does not exist, then $x_0 \in \partial\mathcal{D}_0$. Furthermore, such points cannot exist, if the set $\{f(x,a) \mid a \in A\}$ is convex for all $x \in \mathbb{R}^n$.

We now want to characterize the robust domain of attraction in a way that makes it accessible to theoretic considerations. The idea is to use methods from optimal control. To this end we define a function $g : \mathbb{R}^n \times A \to \mathbb{R}_+$ that associates to each point and perturbation value a certain cost. Then we can define the value function (which may take on the value ∞) $V : \mathbb{R}^n \to \mathbb{R} \cup \{+\infty\}$ by

$$V(x) = \sup_{a \in \mathcal{A}} \int_0^{+\infty} g(\varphi(t;x,a), a(t))\,dt\,, \quad (5)$$

and in order to obtain a function that is finite everywhere we apply the transformation

$$v(x) = 1 - e^{-V(x)}. \quad (6)$$

The idea is now to define the function g and thus the optimal control problem in such a manner, that V, respectively v characterize the robust domain of attraction. The construction needed to do this is simple. The unique feature of the points in the domain of attraction is that all trajectories emanating from these points converge to zero. If $x = 0$ is the only zero of g and g is also bounded away from 0 outside of a bounded neighborhood of 0 then it is clear that for all trajectories that do not converge to zero the integral in (5) will diverge to infinity. If a suitable condition on g in a neighborhood is added then we obtain, that the value of the integral in (5) is finite if and only if the trajectory converges to zero, so that a point x belongs to $\mathcal{D}_0$ if and only if $V(x)$ is finite or equivalently if and only if $v(x) < 1$. We now state the precise assumptions that lead to the desired result.

Assumption 3 *The running cost* $g : \mathbb{R}^n \times A \to \mathbb{R}$ *is continuous and satisfies*

(i) For any $a \in A, g(0,a) = 0$ *and* $g(x,a) > 0$ *for all* $x \neq 0$.

(ii) There exists a constant $g_0 > 0$ *such that* $g(x,a) \ge g_0$ *for all* $a \in A, x_0 \ge r$.

(iii) The function g is locally Lipschitz continuous in x uniformly in a, i.e. for every $R > 0$ *there exists a constant* L_R *such that*

$$\|g(x,a) - g(y,a)\| \le L_R\|x-y\| \text{ for all } \|x\|,\|y\| \le R, \text{ and all } a \in A\,.$$

Since g is nonnegative it is immediate that $V(x) \ge 0$ and $v(x) \in [0,1]$ for all $x \in \mathbb{R}^n$. Under the previous assumption the following result may be shown for the functions V and v.

Proposition 4 ([4]) *Consider system* (3) *and a function* $g : \mathbb{R}^n \times A \to \mathbb{R}$ *such that the Assumptions 1 and 3 are satisfied. Then*

(i) $V(x) < +\infty$ *if and only if* $x \in \mathcal{D}_0$.

(ii) $V(x) = 0$ *if and only if* $x = 0$.

(iii) V *is continuous on* $\mathcal{D}_0$.

(iv) $V(x) \to +\infty$ *for* $x \to x_0 \in \partial\mathcal{D}_0$ *and for* $\|x\| \to \infty$.

(v) $v(x) < 1$ *if and only if* $x \in \mathcal{D}_0$.

(vi) $v(x) = 0$ *if and only if* $x = 0$.

(vii) v *is continuous on* $\mathbb{R}^n$.

(viii) $v(x) \to 1$ *for* $x \to x_0 \in \partial\mathcal{D}_0$ *and for* $\|x\| \to \infty$.

Thus in the construction of v in the manner described above we have found a function that characterizes the robust domain of attraction just as the solution of Zubov's classical equation describes the domain of attraction. What is more is that if we consider the following straightforward generalization of Zubov's equation

$$\inf_{a\in A} \{-Dv(x)f(x,a) - (1 - v(x))g(x,a)\} = 0 \qquad x \in \mathbb{R}^n, \tag{7}$$

then we obtain the following theorem.

Theorem 5 ([4]) *Consider the system* (3) *and a function* $g : \mathbb{R}^n \times A \to \mathbb{R}$ *such that the Assumptions 1 and 3 are satisfied. Then the partial differential equation* (7) *has a unique bounded and continuous viscosity solution* v *on* $\mathbb{R}^n$ *satisfying* $v(0) = 0$.

This solution coincides with v *defined in* (6). *In particular, we have the characterization*

$$\mathcal{D}_0 = \{x \in \mathbb{R}^n \,|\, v(x) < 1\}.$$

A few words of discussion are in order to interpret this result. First, the reader will note that the first order partial differential equation (7) is solved in the viscosity sense. Indeed, an example in [4] shows that classical solutions may not exist so that we have to resort to a solution sense that allows for solutions that are not (everywhere) differentiable. A good introduction to the corresponding theory is provided by the book [1].

Secondly, as in the case of Zubov's original result the importance of the result is twofold. First of all it provides an existence result and then it also provides way of calculating domains of attraction. It has to be mentioned however, that the existence result was already shown to be true in [9, 11]. On the other hand, these papers do not propose methods for the calculation of robust domains of attraction. It will be seen from the examples in the next section that Theorem 5 indeed provides a way for the calculation of domains of attraction. Numerical details of the method are discussed in [2].

3 Examples

In this section we present some examples in order to demonstrate the applicability of the previous results. The examples are in dimension 2 as the representation of the results is especially easy in this case.

Example 6 This example is a modification of Example 4.13 in [8, p. 190]. A perturbation term has been added in one of the nonlinearities. We consider the system

$$\begin{aligned} \dot{x} &= -2x + xy \\ \dot{y} &= -y + (1+a)xy. \end{aligned} \tag{8}$$

Note that the linearization is exponentially stable for any $a \in \mathbb{R}$. Furthermore both the x- and the y-axis, i.e. the sets $\{(x,0) \mid x \in \mathbb{R}\}$ and $\{(0,y) \mid y \in \mathbb{R}\}$, are invariant under the flow and the perturbation does not act on these sets. Thus both x- and y-axis are contained in the robust domain of attraction of the origin no matter what the set of perturbations is as long as it is compact. In Figure 2 the robust domain of attraction corresponding to the set of perturbations $A = [-0.5, 0.5]$ is depicted.

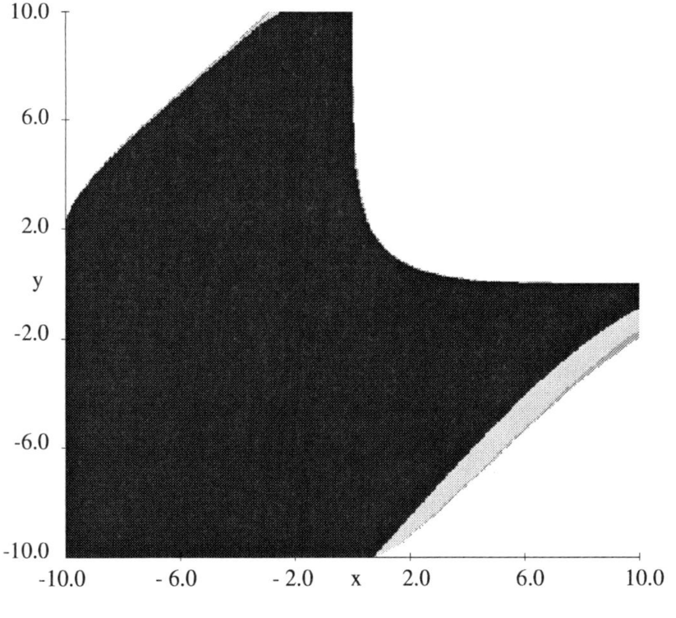

Figure 2: Domain of attraction for system (8) with perturbation range $[-0.5, 0.5]$

The area where the value function is close to 0 is dark grey. The light grey represents regions where intermediate values in the interval $[0.2, 0.95]$ are attained. This corresponds to the fact

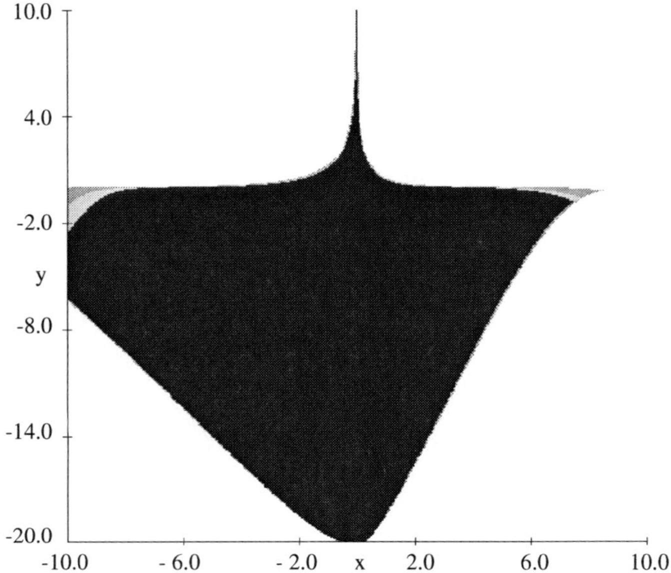

Figure 3: Domain of attraction for system (8) with perturbation range $[-3,3]$

that optimal trajectories take a long time before converging to the origin. The white area is outside of the robust domain of attraction.

Note that to produce Figure 2 the PDE (7) is solved just on the depicted domain. As a consequence trajectories that leave the area are treated as unstable trajectories. In other words, the calculated domain of attraction is the set of all points such that all trajectories converge to the origin *and* do not leave the box $[-10,10]^2$. In fact, we intend to show by a simple argument that the whole half-plane

$$Z := \{[x,y]' \in \mathbb{R}^2 \mid y \leq 0\}$$

is contained in the robust domain of attraction. This may be seen as follows. Consider a compact perturbation set $A \subset \mathbb{R}$ and let $a^+ = \max\{a \in A\}$. We have already noted that the x- and y-axis are invariant under the flow and contained in the domain of attraction. Thus also Z is forward invariant under (8). Consider the Lyapunov function candidate $V(x,y) = x^2 + y^2$. We analyze the behavior of $\dot{V}$ on Z. Denote the right hand side of (8) by $f(x,y,a)$ then we have

$$\dot{V}(x,y,a) = \nabla V(x,y) \cdot f(x,y,a) = 2(y-2)x^2 + 2((1+a)x-1)y^2 .$$

Note that the first term on the right hand side is always negative as $y \leq 0$ in Z. Hence for $\dot{V} < 0$ it is sufficient that $x < (1+a^+)^{-1}$ and so for the set

$$\tilde{Z} := \{[x,y]' \in \mathbb{R}^2 \mid y \leq 0, x < (1+a^+)^{-1}\}$$

we have that $\dot{V}(x,y,a) < 0$ no matter what $a \in A$ we choose. Also it is clear that the set $\tilde{Z}$ is forward invariant, because $\dot{x} < 0$ in $\tilde{Z}$, so that $\tilde{Z}$ is contained in the robust domain of attraction. Now, we have by definition that

$$\dot{x} = -2x + xy = (y-2)x$$

and in Z the factor of x in the right hand side of the equation is negative. This implies that for any initial condition $[x,y]' \in Z$ and any $a(\cdot) \in \mathcal{A}$ the trajectory $\varphi(t;[x,y]',a(\cdot))$ will enter $\tilde{Z}$ in finite time that is bounded independently of $a(\cdot) \in \mathcal{A}$. This shows that all trajectories with initial condition in Z converge to the origin or in other words that $Z \subset \mathcal{D}_0$.

Such a result cannot be obtained by numerical calculations in the approach that has been used, because the calculations are always restricted to bounded domains. So that we may learn from this example, that the numerical results are always to be treated with care. When solving Zubov's equation on a certain domain We do not calculate the robust domain of attraction, but the *relative* robust domain of attraction, that is, the set of points robustly converging to the origin with the additional property that no trajectory leaves the region of calculation.

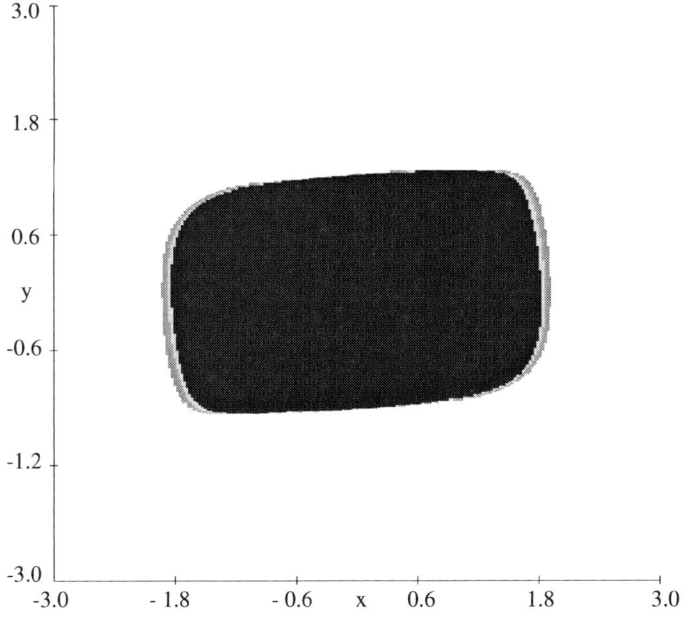

Figure 4: Domain of attraction for system (9) with perturbation range $[-0.1, 0.1]$

In Figure 3 we see the numerical calculation of the robust domain of attraction corresponding to the perturbation set $[-3,3]$. Again it has to be noted that in reality the whole lower half plane is contained in the robust domain of attraction. Also it is interesting to see that the algorithm was not able to detect that the right half of the x-axis is contained in the robust domain of attraction. The reason for this is that trajectories starting close to the x-axis but not on it, move very quickly away from it. So that even when using fine gridding it is hard to see that a part of the dynamics is in fact going to the origin.

Example 7 As a second example consider the system

$$\begin{aligned} \dot{x} &= y \\ \dot{y} &= -x-y+(1+a)(x+2y)(1-y^2), \end{aligned} \tag{9}$$

which is a modification of Exercise 4.17 in [8, p. 199]. Again a perturbation term has been added in one of the nonlinearities.

The calculated domain of attraction for the perturbation set $[-0.1, 0.1]$ is shown in Figure 4. In this case we do not have to worry as in the previous example whether there are points in the robust domain of attraction that are "missed" in our calculations. The reason for this is that the region we have chosen for the calculations contains the calculated robust domain of attraction in its interior. In this case it is shown in [4, Theorem 3.12] that the property $v(x) < 1$ characterizes the robust domain of attraction.

4 Conclusions

We have discussed a method for the calculation of domains of attraction for perturbed systems. The method relies on the ability to numerically solve for a viscosity solution of a first order PDE. This has been an active research area over the past decade and quite efficient methods are available, see for instance the appendix by M. Falcone in [1]. For lower dimensional systems this provides an efficient method to obtain estimates for domains of attraction. As we have seen any calculation on a bounded domain just leads to the calculation of a relative domain of attraction but this may also be interpreted as a merit of the method, because in many cases one is not interested in solutions that move to far away from the equilibrium position.

It is clear that the approach is restricted to low dimensions because for the calculations a region of the state space has to be gridded which requires exponential cost in the dimension of the problem. However, to the best of our knowledge this is the first possibility to calculate domains of attractions for systems subject to perturbations available in the literature. Other popular methods based on Lyapunov functions clearly give lower estimates, but provide no method for checking how conservative these estimates are. Which of the methods is more appropriate would therefore seem to depend on the problem under investigation and the desired result.

5 Acknowledgement

The author wants to thank Fabio Camilli and Lars Grüne for a fruitful collaboration on the subject of Zubov's method and numerous valuable discussions. The examples have been calculated using the program developed by Lars Grüne.

References

[1] M. Bardi and I. Capuzzo Dolcetta, *Optimal Control and Viscosity Solutions of Hamilton-Jacobi-Bellman equations*, Birkhäuser, Boston, 1997.

[2] F. Camilli, L. Grüne, and F. Wirth. A regularization of Zubov's equation for robust domains of attraction. in: *Nonlinear Control in the Year 2000*, A. Isidori et al., eds., Lecture Notes in Control and Information Sciences, Vol. 258, pp. 277 - 290, Springer-Verlag, London, 2000.

[3] F. Camilli, L. Grüne, and F. Wirth, Zubov's method for perturbed differential equations, Proc. Symposium on Mathematical Theory of Networks and Systems MTNS-2000, Perpignan, France, July 2000. (CD-Rom) paper B100.

[4] F. Camilli, L. Grüne, and F. Wirth, A generalization of Zubov's method to perturbed systems, *SIAM J. Control Optimization* 40(2):496–515, 2002.

[5] F. Camilli, L. Grüne, F. Wirth. A generalization of Zubov's method to perturbed systems. Proc. 41st IEEE Conference on Decision and Control, CDC2002, Las Vegas, NV, US, December 2002.

[6] 0. Hachicho, B. Tibken. Estimating domains of attraction of a class of nonlinear dynamical systems with LMI methods based on the theory of moments. Proc. 41st IEEE Conference on Decision and Control (CDC 2002), Las Vegas, NV, December 2002, pp. 3150–3155.

[7] W. Hahn, *Stability of Motion*, Springer-Verlag, Berlin, 1967.

[8] H. K. Khalil. *Nonlinear Systems*. 2nd ed. Prentice-Hall, Upper Saddle River, NJ, 1996.

[9] Y. Lin, E. D. Sontag, and Y. Wang. A smooth converse Lyapunov theorem for robust stability. *SIAM J. Control Optim.* 34:124–160, 1996.

[10] A. Papachristodoulou, S. Prajna. On the construction of Lyapunov functions using the sum of squares decomposition. Proc. 41st IEEE Conference on Decision and Control (CDC 2002), Las Vegas, NV, December 2002, pp. 3482–3487.

[11] L. Praly and A. Teel. A smooth Lyapunov function from a class-$\mathcal{KL}$ estimate involving two positive semidefinite functions. ESAIM Control, Optimization and Calculus of Variations 5:313-367, 2000.

[12] B. Tibken, K.F. Dilaver. Computation of subsets of the domain of attraction for polynomial systems. Proc. 41st IEEE Conference on Decision and Control (CDC 2002), Las Vegas, NV, December 2002, pp. 2651–2656.

[13] A. Vannelli and M. Vidyasagar, Maximal Lyapunov functions and domains of attraction for autonomous nonlinear systems. *Automatica* 21(1):69–80, 1985.

[14] F.W. Wilson, The structure of the level surfaces of a Lyapunov function. *J. Diff. Equations* 3:323–329, 1967.

[15] V.I. Zubov. *Methods of A.M. Lyapunov and their Application*, P. Noordhoff, Groningen, 1964.

Parameter- und Zustandsschätzung mit Zustandsschätzern auf bewegtem Horizont

Jens Vortisch

Institut für Automatisierungstechnik, Universität Bremen, FB1/NW1, Otto-Hahn-Allee, 28359 Bremen
Email: vortisch@iat.uni-bremen.de

Zusammenfassung

Bei der Systemüberwachung und der Fehlerdiagnose tritt oft das Problem auf, dass nicht nur die Zustandsgrössen eines Systems sondern auch bestimmte Modellparameter geschätzt werden müssen. In diesem Artikel werden zwei Zustandsschätzeransätze für die Parameter- und Zustandsschätzung asynchron und nichtäquidistant abgetasteter nichtlinearer Systeme vorgestellt. Dabei handelt es sich um sogenannte Zustandsschätzer auf bewegtem Horizont, die auf der Echtzeitoptimierung quadratischer Gütemasse mit Hilfe des Gauss-Newton-Verfahrens basieren. Am Beispiel des Versuchsaufbaus Dreitanksystem werden die vorgestellten Beobachtungsverfahren auf ihre Eigenschaften hinsichtlich der Konvergenz und dem notwendigem Rechenaufwand untersucht und verglichen.

1 Einleitung

Die Beschreibung eines technischen Systems kann oft durch ein nichtlineares Zustandsraummodell erfolgen. Durch die Anwendung physikalischer Gesetzmäßigkeiten wird ein System gewöhnlicher nichtlinearer Differentialgleichungen erster Ordnung aufgestellt. Die Parameter dieses Differentialgleichungssystems können nicht immer allein durch theoretische Betrachtungen bestimmt werden. Stattdessen werden Parameterschätzverfahren (z.B. [3], [4]) eingesetzt, die aus gemessenen Ein- und Ausgangsgrößen die Werte der Modellparameter ermitteln.
Während ein Zustandsraummodell eine zeitkontinuierliche Beschreibung darstellt, werden die Ausgangsgrößen in der Regel zeitdiskret gemessen. In diesem Artikel wird von einer asynchronen und nichtäquidistanten Abtastung ausgegangen. Bei einem System, das mehrere Ausgangsgrößen besitzt, werden die verschiedenen Ausgangsgrößen dabei nicht gleichzeitig (asynchron) und mit beliebiger, variabler Abtastzeit (nichtäquidistant) gemessen.
Die Parameterschätzung für ein derartiges Zustandsraummodell ist schwierig, weil sie auf das Problem der gleichzeitigen Schätzung von Zuständen und Systemparametern führt. Mit der Zustandserweiterung und der adaptiven nichtlinearen Filterung stehen dafür zwei Methoden zur Verfügung. Bei der Zustandserweiterung werden die Systemparameter als Zustände aufgefasst und mit einem nichtlinearen Zustandsschätzer geschätzt (z.B. [5]). Bei der adaptiven Filterung wird ein nichtlineares Filter so eingestellt, dass das Verhalten

des optimal eingestellten Filters dem wirklichen Systemverhalten entspricht (z.B. [1]). Dem Systemmodell des Filters entnimmt man die Modellparameter. Für die Adaption der Filterparameter wird ein Gütemaß aufgestellt, das im Echtzeitbetrieb rekursiv minimiert wird.
In diesem Artikel werden die beiden oben genannten Methoden der Parameter- und Zustandsschätzung anhand zweier eng verwandter Zustandsschätzeransätze auf bewegtem Horizont verglichen. Die Parameterschätzung durch Zustandserweiterung wird dabei mit Hilfe des sogenannten *Gauss-Newton Moving Horizon Observer* [7] vorgenommen. Eine Erweiterung des *GNMHO*, die in veränderter Form auch zur Schätzung von Systemstörungen benutzt wird, kann als ein wie oben beschriebener adaptiver Filter verstanden werden. Beiden Verfahren ist gemein, dass ein, aus einer Reihe von Messwerten (dem sogenannten Beobachtungshorizont) gebildetes, quadratisches Gütemaß mit Hilfe des Gauss-Newton-Verfahrens minimiert und so die Konvergenz der Schätzwerte hergestellt wird.
Beide Verfahren stellen einen Spezialfall einer allgemeinen Klasse von nichtlinearen Zustandsschätzern auf bewegtem Horizont (*Moving Horizon Estimator*) dar. Eine umfassende Übersicht hierzu bieten [6] und [2].
In Abschnitt 2 wird zunächst der *GNMHO* und das Verfahren der Parameterschätzung durch Zustandserweiterung vorgestellt. Daran anschließend ist in Abschnitt 3 die Erweiterung des *GNMHO* zur gleichzeitigen Parameter- und Zustandsbeobachtung beschrieben. Die Anwendung der Verfahren durch Simulation und an einem realen System und die Diskussion der Ergebnisse folgt in Abschnitt 4. Abschließend erfolgt eine kurze Zusammenfassung mit einer Wertung der erzielten Ergebnisse.

2 Der Gauss-Newton Moving Horizon Observer

2.1 Das nichtlineare Modell der Strecke

Die Klasse der betrachteten nichtlinearen Systeme kann durch das folgende zeitkontinuierliche nichtlineare Modell mit Störungen in den Messgleichungen beschrieben werden:

$$\begin{aligned} \dot{\mathbf{x}} &= \mathbf{f}(\mathbf{x},\mathbf{u}), \\ \mathbf{x}(0) &= \mathbf{x}_0, \\ \mathbf{y} &= \mathbf{h}(\mathbf{x},\mathbf{u}) + \mathbf{v}. \end{aligned} \tag{1}$$

Es besteht aus einem nichtlinearen gewöhnlichen Differentialgleichungssystem und einer Messgleichung, wobei $\mathbf{x}$ den Zustandsvektor der Ordnung n und $\mathbf{u}$ den bekannten Eingangsvektor der Ordnung p darstellt. In der Messgleichung ist $\mathbf{y}$ der Messvektor und $\mathbf{v}$ ein Störungsvektor, der Messrauschen oder eine ungenügende Modellierung des Messgliedes repräsentiert. Beide sind von der Ordnung q.
Da in diesem Artikel das Problem der Zustandsschätzung von asynchron und nichtäquidistant abgetasteten nichtlinearen Systemen behandelt wird, wird zunächst die zeitkontinuierliche Darstellung des nichtlinearen Systems (1) in eine zeitdiskrete Darstellung umgeformt. Dies geschieht, wie bei linearen Systemen, durch das Lösen des nichtlinearen

Differentialgleichungssystems:

$$
\begin{aligned}
\mathbf{x}_{k+1} &= \mathbf{x}_k + \int_{t_k}^{t_{k+1}} \mathbf{f}(\mathbf{x}(t), \mathbf{u}(t))\, dt, \\
\mathbf{x}(0) &= \mathbf{x}_0, \\
\mathbf{y_k} &= \mathbf{h}(\mathbf{x}_k, \mathbf{u}_k) + \mathbf{v}_k.
\end{aligned} \tag{2}
$$

Im Allgemeinen kann das Integral aus (2) nicht geschlossen dargestellt werden, sondern muss mit Hilfe numerischer Verfahren wie z.B. dem Runge-Kutta-Verfahren berechnet werden. Zur Vereinfachung der Schreibweise kann (2) trotzdem als:

$$
\begin{aligned}
\mathbf{x}_{k+1} &= \mathbf{F}(\mathbf{x}_k, \mathbf{u}_k), \\
\mathbf{x}(0) &= \mathbf{x}_0, \\
\mathbf{y_k} &= \mathbf{h}(\mathbf{x}_k, \mathbf{u}_k) + \mathbf{v}_k
\end{aligned} \tag{3}
$$

dargestellt werden, wenn man annimmt, dass $\mathbf{u}(t)$ während eines Abtastintervalls konstant bleibt.

2.2 Aufbau des Zustandsschätzers

Für das in Abschnitt 1 beschriebene Problem der Zustandsbeobachtung asynchron und nichtäquidistant abgetasteter nichtlinearer Systeme wird der Ansatz von [8] als Ausgangspunkt gewählt. Zum einen weist er durch die Anwendung des Newton-Verfahrens ein akzeptables Verhältnis von Rechenaufwand und Konvergenz auf, zum anderen lässt sich der Rechenaufwand weiter verringern, weil das Newton-Verfahren für Quadratsummenfunktionen als Gütemaß durch ein Näherungsverfahren, das rechnerisch weit weniger aufwendige Gauss-Newton-Verfahren, approximiert werden kann.
Das Verfahren wird der Einfachheit halber für nichtlineare Systeme mit nur einem Ausgang dargestellt:

$$
\begin{aligned}
\mathbf{x}_{k+1} &= \mathbf{F}(\mathbf{x}_k, \mathbf{u}_k), \\
\mathbf{x}(0) &= \mathbf{x}_0, \\
y_k &= h(\mathbf{x}_k, \mathbf{u}_k) + v_k.
\end{aligned} \tag{4}
$$

Das zu minimierende Gütemaß lautet:

$$
\min_{\hat{\mathbf{x}}_{k-N|k}} \quad \Phi_k = \sum_{i=k-N}^{k} \hat{v}_{i|k} R^{-1} \hat{v}_{i|k}. \tag{5}
$$

Mit den nichtlinearen Gleichungsnebenbedingungen:

$$
\begin{aligned}
\hat{\mathbf{x}}_{j+1|k} &= \mathbf{F}(\hat{\mathbf{x}}_{j|k}, \mathbf{u}_j) \qquad &, j = k-N, \ldots, k-1, \\
\hat{v}_{j|k} &= y_k - h(\hat{\mathbf{x}}_{j|k}, \mathbf{u}_k) \qquad &, j = k-N, \ldots, k.
\end{aligned} \tag{6}
$$

Die Notation $\hat{\mathbf{x}}_{i|j}$ bezeichnet die Schätzung von $\mathbf{x}$ zur Zeit i, die sich auf die Messwerte bis zur Zeit j stützt. Dasselbe gilt für die Schätzung der Messstörungen $\hat{v}_{i|j}$.
Zunächst wird das Gütemaß in zu (5) leicht veränderter Gestalt angegeben:

$$
\min_{\hat{\mathbf{x}}_{k-N|k}} \quad \Phi_k = \hat{\mathbf{v}}^{\mathrm{T}} \mathbf{W} \hat{\mathbf{v}}. \tag{7}
$$

Der Vektor $\hat{\mathbf{v}}$ wird aus allen innerhalb des Beobachtungshorizonts liegenden Fehlern zwischen Mess- und Schätzwerten $\hat{v}_{i|k},\quad k-N \leq i \leq k$ zusammengesetzt (5):

$$\hat{\mathbf{v}} \;=\; \begin{pmatrix} \hat{v}_{k-N|k} \\ \vdots \\ \hat{v}_{k|k} \end{pmatrix}. \tag{8}$$

Die Gewichtungsmatrix $\mathbf{W}$ lautet dann (5):

$$\mathbf{W} \;=\; R^{-1}\mathbf{I}. \tag{9}$$

Die erste und zweite Ableitung von Φ_k nach dem Anfangswert $\hat{\mathbf{x}}_{k-N|k}$ auf dem Beobachtungshorizont lautet:

$$\frac{\partial\, \Phi_k}{\partial\, \hat{\mathbf{x}}_{k-N|k}} \;=\; \mathbf{J}^{\mathrm{T}}(\mathbf{W}^{\mathrm{T}}+\mathbf{W})\hat{\mathbf{v}}, \tag{10}$$

$$\frac{\partial^2\, \Phi_k}{\partial\, \hat{\mathbf{x}}^2_{k-N|k}} \;=\; \mathbf{J}^{\mathrm{T}}(\mathbf{W}^{\mathrm{T}}+\mathbf{W})\mathbf{J} + \underbrace{\sum_{i=k-N}^{k} \hat{\mathbf{v}}^{\mathrm{T}}(\mathbf{W}^{\mathrm{T}}+\mathbf{W})_i\mathbf{H}_i}_{\approx 0}, \tag{11}$$

wobei $\mathbf{J}$ die Jakobi-Matrix von $\hat{\mathbf{v}}$ und $\mathbf{H}_i$ die Hesse'sche Matrix der i-ten Zeile von $\hat{\mathbf{v}}$ darstellt. $(\mathbf{W}^{\mathrm{T}}+\mathbf{W})_i$ bezeichnet dabei die i-te Spalte von $(\mathbf{W}^{\mathrm{T}}+\mathbf{W})$. Wenn die Fehler zwischen den Messwerten und den Schätzungen klein genug sind, können diese Hesse'schen Matrizen vernachlässigt werden, so dass eine Näherung für die zweite Ableitung des Gütemaßes angegeben werden kann, die nur von der ersten Ableitung des Gütemaßes abhängt. Die ansonsten rechentechnisch sehr aufwendige numerische Berechnung der zweiten Ableitung kann damit stark vereinfacht werden.
Damit Φ_k minimal wird, muss (10) notwendigerweise gleich dem Nullvektor sein. Eine Näherung von (10) durch eine nach dem linearen Glied abgebrochene Taylor-Reihe ergibt:

$$\frac{\partial\, \Phi_k}{\partial\, \hat{\mathbf{x}}_{k-N|k}} + \frac{\partial^2\, \Phi_k}{\partial\, \hat{\mathbf{x}}^2_{k-N|k}}\Delta\hat{\mathbf{x}}_{k-N|k} \;=\; \mathbf{0}. \tag{12}$$

Die Gleichung (12) ergibt zusammen mit (11) und (10) die Vorschrift für die Durchführung eines Gauss-Newton-Schrittes:

$$\Delta\hat{\mathbf{x}}_{k-N|k} \;=\; -\left[\mathbf{J}^{\mathrm{T}}(\mathbf{W}^{\mathrm{T}}+\mathbf{W})\mathbf{J}\right]^{-1}\mathbf{J}^{\mathrm{T}}(\mathbf{W}^{\mathrm{T}}+\mathbf{W})\hat{\mathbf{v}}. \tag{13}$$

Die Anwendung von (13) zur Berechnung einer verbesserten Schätzung des Anfangswerts lautet nun folgendermaßen:

$$\hat{\mathbf{x}}^+_{k-N|k} \;=\; \hat{\mathbf{x}}^-_{k-N|k} + \Delta\hat{\mathbf{x}}^-_{k-N|k}. \tag{14}$$

Durch Simulation des Parallelmodells wird ausgehend von dem verbesserten geschätzten Anfangswert $\hat{\mathbf{x}}^+_{k-N|k}$ der geschätzte Anfangswert für das um einen Messwert vorangeschrittene Beobachtungsfenster bestimmt:

$$\hat{\mathbf{x}}^-_{k-N+1|k+1} \;=\; \hat{\mathbf{x}}^+_{k-N+1|k}. \tag{15}$$

Die Indizes + und - bezeichnen dabei jeweils die Schätzwerte nach bzw. vor der Durchführung des Gauss-Newton-Schrittes.
Die zur Berechnung eines Gauss-Newton-Schrittes notwendige Jakobi-Matrix des Messfehlervektors, der aus $\hat{v}_{i|k}, \quad k-N \leq i \leq k$ zusammengesetzt ist, wird wie folgt gebildet:

$$\mathbf{J} \; = \; \begin{pmatrix} \left(-\frac{\partial\, h(\hat{\mathbf{x}}_{k-N|k})}{\partial\, \hat{\mathbf{x}}}\right)^{\mathrm{T}} \frac{\partial\, \hat{\mathbf{x}}_{k-N|k}}{\partial\, \hat{\mathbf{x}}_{k-N|k}} \\ \vdots \\ \left(-\frac{\partial\, h(\hat{\mathbf{x}}_{k|k})}{\partial\, \hat{\mathbf{x}}}\right)^{\mathrm{T}} \frac{\partial\, \hat{\mathbf{x}}_{k|k}}{\partial\, \hat{\mathbf{x}}_{k-N|k}} \end{pmatrix}. \tag{16}$$

Während die Ableitungen der Ausgangsfunktionen nach den Schätzwerten $\partial\, h(\hat{\mathbf{x}}_{k-i|k})/\partial\, \hat{\mathbf{x}}$ leicht analytisch gebildet werden können, erfordert die Berechnung von
$\mathbf{X}(t_i) = \partial\, \hat{\mathbf{x}}_{i|k}/\partial\, \hat{\mathbf{x}}_{k-N|k}$ die Lösung des matrixwertigen Anfangswertproblems (17) in jedem Zeitpunkt, an dem eine Messung aufgenommen wurde,

$$\begin{aligned} \dot{\mathbf{X}} &= \frac{\partial\, \mathbf{f}(\hat{\mathbf{x}})}{\partial\, \hat{\mathbf{x}}}\mathbf{X}, \\ \mathbf{X}(0) &= \mathbf{I}. \end{aligned} \tag{17}$$

Durch die Gleichungen (6)-(9) und (13)-(17) ist der *GNMHO* vollständig definiert.
Für Systeme mit mehreren Messgrößen ist die Gewichtungsmatrix $\mathbf{W}$ nicht zeitinvariant, wie in Gleichung (9) angegeben wurde. Weil der Vektor der in dem Beobachtungshorizont liegenden $N+1$ Messgrößen wegen der asynchronen Messwertaufnahme aus Messwerten verschiedener Ausgänge zusammengesetzt werden kann, ändert sich die Zusammensetzung der Matrix $\mathbf{W}$ entsprechend.
Die Zahl der Gauss-Newton-Schritte, die von dem Schätzalgorithmus ausgeführt werden, ist nur von der für die Echtzeitberechnung zur Verfügung stehenden Zeit abhängig. Diese ist wegen des asynchronen und nichtäquidistanten Charakters der Messwertaufnahme im voraus unbekannt. Deshalb ist es sinnvoll, so viele Iterationen des Gauss-Newton-Verfahrens durchzuführen wie möglich, bis ein neuer Messwert aufgenommen wird und sich der Beobachtungshorizont deshalb verschiebt.
Voraussetzung für die eindeutige Lösung des Gleichungssystems (12) ist, dass $\mathbf{J}^{\mathrm{T}}\mathbf{J}$ nicht singulär ist, bzw. $\mathbf{J}$ einen Rang von n hat. Dann hat das Verfahren, wenn tatsächlich $\hat{\mathbf{v}} = \mathbf{0}$ gilt, quadratische Konvergenz wie das Newton-Verfahren, obwohl nur die ersten Ableitungen berechnet werden. Sind diese Voraussetzungen nicht erfüllt, so konvergiert das Verfahren langsamer oder gar nicht. Für diesen Fall sind aber verschiedene alternative Verfahren (z. B. das Levenberg-Marquardt-Verfahren) bekannt.
Selbst wenn die Jakobi-Matrix $\mathbf{J}$ den vollen Rang hat, führt ein Gauss-Newton-Schritt nicht notwendigerweise zu einer Abnahme des Gütemaßes (7), denn die notwendige Bedingung für ein Minimum (12) ist wie aus der Herleitung ersichtlich nur eine Näherung der wahren notwendigen Bedingung. Um eine Abnahme des Gütemaßes zu bewirken, setzt man deshalb eine Schrittweitensteuerung ein. Die Schrittweite wird in (14) durch einen Faktor α eingestellt:

$$\hat{\mathbf{x}}^{+}_{k-N|k} \; = \; \hat{\mathbf{x}}^{-}_{k-N|k} + \alpha \Delta \hat{\mathbf{x}}^{-}_{k-N|k}. \tag{18}$$

Beim *GNMHO* ist die Komplexität der notwendigen Berechnungen beträchtlich. Dies liegt vor allem an der numerischen Integration von $n+n^2$ verkoppelten nichtlinearen

Differentialgleichungen, die zur Ausführung jedes Gauss-Newton-Schrittes durchgeführt werden müssen. Die Zahl $n + n^2$ setzt sich zum einen aus den n nichtlinearen Differentialgleichungen des Parallelmodells und zum anderen aus den n^2 nichtlinearen Differentialgleichungen aus (17) zusammen. Diese $n + n^2$ nichtlinearen Differentialgleichungen müssen über den gesamten Beobachtungshorizont von $N + 1$ Messwerten numerisch gelöst werden. Außerdem kann es sein, dass, wenn ein Gauss-Newton-Schritt nicht sofort zur Abnahme des Gütemaßes führt, eine unbestimmte Anzahl der $n + n^2$ Simulationen der Differentialgleichungen ausgeführt werden müssen bis eine geeignete Schrittweite α für den Gauss-Newton-Schritt (18) gefunden wird.

2.3 Beobachtung von Systemparametern durch Zustandserweiterung

Besitzt ein nichtlineares System r unbekannte oder nur näherungsweise bekannte Parameter, so kann man diese in einem Vektor $\mathbf{a}$ zusammenfassen. Das System kann dann in der Form

$$\begin{aligned} \dot{\mathbf{x}} &= \mathbf{f}(\mathbf{x}, \mathbf{a}, \mathbf{u}), \\ \mathbf{x}(0) &= \mathbf{x}_0, \\ \mathbf{y} &= \mathbf{h}(\mathbf{x}, \mathbf{a}, \mathbf{u}) \end{aligned} \tag{19}$$

angegeben werden. Unter der Voraussetzung, dass der Parametervektor $\mathbf{a}$ konstant ist, wird nun der Zustandsvektor $\mathbf{x}$ folgendermaßen erweitert:

$$\mathbf{x}_{erw} := \begin{pmatrix} \mathbf{x} \\ \mathbf{a} \end{pmatrix}. \tag{20}$$

Für die Zustandsraumdarstellung des nichtlinearen Systems gilt dann:

$$\begin{aligned} \dot{\mathbf{x}}_{erw} &= \mathbf{f}_{erw}(\mathbf{x}_{erw}, \mathbf{u}) = \begin{pmatrix} \mathbf{f}(\mathbf{x}_{erw}, \mathbf{u}) \\ \mathbf{0} \end{pmatrix}, \\ \mathbf{x}_{\mathbf{erw}}(\mathbf{0}) &= \begin{pmatrix} \mathbf{x}_0 \\ \mathbf{a}_0 \end{pmatrix}, \\ \mathbf{y} &= \mathbf{h}(\mathbf{x}_{erw}, \mathbf{u}). \end{aligned} \tag{21}$$

Der Erweiterung des Systems durch Parameter sind aber Grenzen gesetzt. Das nichtlineare System kann nämlich nur um so viele Zustände erweitert werden wie Messgleichungen existieren, d.h. die Zahl r der zu beobachtenden Parameter darf höchstens so groß sein wie die Zahl der q Messgleichungen. Ein Beweis dafür kann in [8] nachgelesen werden. Entwirft man nun einen Beobachter auf bewegtem Horizont für das erweiterte System, wie in Abschnitt 2.2 beschrieben, so ist zur Beobachtung die Lösung von $n+r+(n+r)^2$ Differentialgleichungen erforderlich.

3 GNMHO mit erweitertem Gütemaß

Eine andere Möglichkeit zur Beobachtung unbekannter Parameter für ein System der Form (19) bietet eine modifizierte Form des *GNMHO*, der in Abschnitt 2 beschrieben wurde.

Ein System der Form (19), wegen der besseren Übersichtlichkeit hier nur mit einer Ausgangsgröße, kann in zeitdiskretisierter Form folgendermaßen dargestellt werden:

$$\begin{aligned} \mathbf{x}_{k+1} &= \mathbf{F}(\mathbf{x}_k, \mathbf{a}, \mathbf{u}_k), \\ \mathbf{x}(0) &= \mathbf{x}_0, \\ y_k &= h(\mathbf{x}_k, \mathbf{a}, \mathbf{u}_k) + v_k. \end{aligned} \tag{22}$$

Das Optimierungsproblem lautet dann:

$$\min_{\hat{\mathbf{x}}_{k-N|k}, \hat{\mathbf{a}}} \quad \Phi_k = \hat{\mathbf{a}}^{\mathrm{T}} \mathbf{Q}^{-1} \hat{\mathbf{a}} + \sum_{i=k-N+1}^{k} \hat{v}_{i|k} R^{-1} \hat{v}_{i|k}, \tag{23}$$

unter Berücksichtigung der nichtlinearen Gleichungsnebenbedingungen:

$$\begin{aligned} \hat{\mathbf{x}}_{j+1|k} &= \mathbf{F}(\hat{\mathbf{x}}_{j|k}, \hat{\mathbf{a}}, \mathbf{u}_j), \qquad & j = k-N, \ldots, k-1, \\ \hat{v}_{j|k} &= y_k - h(\hat{\mathbf{x}}_{j|k}, \hat{\mathbf{a}}, \mathbf{u}_j), \qquad & j = k-N, \ldots, k. \end{aligned} \tag{24}$$

Das Gütemaß enthält nun neben den Schätzfehlern auch die r Parameterschätzungen $\hat{a}_i \;, i = 1, \ldots, r$. Schreibt man das Gütemaß (23) als Produkt:

$$\Phi_k = \begin{pmatrix} \hat{\mathbf{v}} \\ \hat{\mathbf{a}} \end{pmatrix}^{\mathrm{T}} \mathbf{W} \begin{pmatrix} \hat{\mathbf{v}} \\ \hat{\mathbf{a}} \end{pmatrix}, \tag{25}$$

so setzt sich der erweiterte Fehlervektor aus dem Vektor $\hat{\mathbf{v}}$ der Schätzfehler und dem Vektor $\hat{\mathbf{a}}$ der Parameterschätzungen zusammen:

$$\begin{pmatrix} \hat{\mathbf{v}} \\ \hat{\mathbf{a}} \end{pmatrix} = \begin{pmatrix} \hat{v}_{k-N|k} \\ \vdots \\ \hat{v}_{k|k} \\ \hat{a}_1 \\ \vdots \\ \hat{a}_r \end{pmatrix}. \tag{26}$$

Die Gewichtungsmatrizen $\mathbf{R}^{-1}$:

$$\mathbf{R}^{-1} = R^{-1} \mathbf{I} \tag{27}$$

und $\mathbf{Q}^{-1}$ werden zu einer Hypermatrix $\mathbf{W}$ zusammengefasst:

$$\mathbf{W} = \begin{pmatrix} \mathbf{R}^{-1} & 0 \\ 0 & \mathbf{Q}^{-1} \end{pmatrix}. \tag{28}$$

Die Ableitungen des Gütemaßes Φ_k nach dem Anfangswert $\hat{\mathbf{x}}_{k-N|k}$ innerhalb des berücksichtigten Beobachtungshorizonts und dem Vektor $\hat{\mathbf{a}}$ der Parameterschätzungen lauten:

$$\frac{\partial \, \Phi_k}{\partial \begin{bmatrix} \hat{\mathbf{x}}_{k-N|k} \\ \hat{\mathbf{a}} \end{bmatrix}} = \mathbf{J}^{\mathrm{T}} (\mathbf{W}^{\mathrm{T}} + \mathbf{W}) \begin{pmatrix} \hat{\mathbf{v}} \\ \hat{\mathbf{a}} \end{pmatrix}, \tag{29}$$

und

$$\frac{\partial^2 \Phi_k}{\partial \begin{bmatrix} \hat{\mathbf{x}}_{k-N|k} \\ \hat{\mathbf{a}} \end{bmatrix}^2} = \mathbf{J}^{\mathrm{T}}(\mathbf{W}^{\mathrm{T}} + \mathbf{W})\mathbf{J} + \underbrace{\sum_{i=1}^{N+1} \begin{pmatrix} \hat{\mathbf{v}} \\ \hat{\mathbf{a}} \end{pmatrix}_i (\mathbf{W}^{\mathrm{T}} + \mathbf{W})_i \mathbf{H}_i}_{\approx 0}. \tag{30}$$

Sind die Elemente des erweiterten Fehlervektors $\begin{pmatrix} \hat{\mathbf{v}} \\ \hat{\mathbf{a}} \end{pmatrix}_i$ klein genug, so kann die zweite Ableitung des Gütemaßes durch die Jakobi-Matrix des erweiterten Fehlervektors $\mathbf{J}$ angenähert werden. Die Hesse'schen Matrizen $\mathbf{H}_i$ der Elemente des erweiterten Fehlervektors werden deshalb nicht benötigt. Diese Voraussetzung ist im Zusammenhang mit der Parameterschätzung natürlich nicht immer erfüllt. Man kann aber immer durch geeignete Skalierung für ausreichend kleine Parameterwerte und damit kleine Schätzwerte sorgen. Ein Gauss-Newton-Schritt zur Verbesserung des Anfangswerts und des Zustandsstörungsvektors lautet deshalb:

$$\Delta \begin{bmatrix} \hat{\mathbf{x}}_{k-N|k} \\ \hat{\mathbf{a}} \end{bmatrix} = -\left[\mathbf{J}^{\mathrm{T}}(\mathbf{W}^{\mathrm{T}} + \mathbf{W})\mathbf{J}\right]^{-1} \mathbf{J}^{\mathrm{T}}(\mathbf{W}^{\mathrm{T}} + \mathbf{W}) \begin{pmatrix} \hat{\mathbf{v}} \\ \hat{\mathbf{a}} \end{pmatrix}. \tag{31}$$

Die Jakobi-Matrix heißt allgemein:

$$\mathbf{J} = \begin{pmatrix} \frac{\partial \hat{\mathbf{v}}}{\partial \hat{\mathbf{x}}_{k-N|k}} & \frac{\partial \hat{\mathbf{v}}}{\partial \hat{\mathbf{a}}} \\ \frac{\partial \hat{\mathbf{a}}}{\partial \hat{\mathbf{x}}_{k-N|k}} & \frac{\partial \hat{\mathbf{a}}}{\partial \hat{\mathbf{a}}} \end{pmatrix}. \tag{32}$$

Sie kann wie folgt vereinfacht werden:

$$\mathbf{J} = \begin{pmatrix} \frac{\partial \hat{\mathbf{v}}}{\partial \hat{\mathbf{x}}_{k-N|k}} & \frac{\partial \hat{\mathbf{v}}}{\partial \hat{\mathbf{a}}} \\ \mathbf{0} & \mathbf{I} \end{pmatrix}. \tag{33}$$

Während die Schätzfehler über die Systemgleichungen (24) vom geschätzten Anfangswert und den Schätzungen der Parametervektoren abhängen, sind die Schätzungen der Parametervektoren unabhängig vom geschätzten Anfangswert, denn sie werden wie der geschätzte Anfangswert vorgegeben, sind also Optimierungsvariablen. Die Ableitung der Schätzungen der Parametervektoren nach sich selbst ist die Einheitsmatrix $\mathbf{I}$.
Setzt man (33) in (31) ein, so folgt für einen Gauss-Newton-Schritt:

$$\begin{aligned} \Delta \begin{bmatrix} \hat{\mathbf{x}}_{k-N|k} \\ \hat{\mathbf{a}} \end{bmatrix} = & -\begin{bmatrix} \left(\frac{\partial \hat{\mathbf{v}}}{\partial \hat{\mathbf{x}}_{k-N|k}}\right)^{\mathrm{T}} \mathbf{R}^{-1} \frac{\partial \hat{\mathbf{v}}}{\partial \hat{\mathbf{x}}_{k-N|k}} & \left(\frac{\partial \hat{\mathbf{v}}}{\partial \hat{\mathbf{x}}_{k-N|k}}\right)^{\mathrm{T}} \mathbf{R}^{-1} \frac{\partial \hat{\mathbf{v}}}{\partial \hat{\mathbf{a}}} \\ \left(\frac{\partial \hat{\mathbf{v}}}{\partial \hat{\mathbf{a}}}\right)^{\mathrm{T}} \mathbf{R}^{-1} \frac{\partial \hat{\mathbf{v}}}{\partial \hat{\mathbf{x}}_{k-N|k}} & \left(\frac{\partial \hat{\mathbf{v}}}{\partial \hat{\mathbf{a}}}\right)^{\mathrm{T}} \mathbf{R}^{-1} \frac{\partial \hat{\mathbf{v}}}{\partial \hat{\mathbf{a}}} + \mathbf{Q}^{-1} \end{bmatrix}^{-1} \cdot \\ & \cdot \begin{bmatrix} \left(\frac{\partial \hat{\mathbf{v}}}{\partial \hat{\mathbf{x}}_{k-N|k}}\right)^{\mathrm{T}} \mathbf{R}^{-1} \hat{\mathbf{v}} \\ \left(\frac{\partial \hat{\mathbf{v}}}{\partial \hat{\mathbf{a}}}\right)^{\mathrm{T}} \mathbf{R}^{-1} \hat{\mathbf{v}} + \mathbf{Q}^{-1} \hat{\mathbf{a}} \end{bmatrix}. \end{aligned} \tag{34}$$

Die zwei unbekannten Elemente der Gleichung (33) lauten:

$$\frac{\partial \hat{\mathbf{v}}}{\partial \hat{\mathbf{x}}_{k-N|k}} = \overbrace{\begin{pmatrix} \left(-\frac{\partial h(\hat{\mathbf{x}}_{k-N|k})}{\partial \hat{\mathbf{x}}}\right)^{\mathrm{T}} \frac{\partial \hat{\mathbf{x}}_{k-N|k}}{\partial \hat{\mathbf{x}}_{k-N|k}} \\ \vdots \\ \left(-\frac{\partial h(\hat{\mathbf{x}}_{k|k})}{\partial \hat{\mathbf{x}}}\right)^{\mathrm{T}} \frac{\partial \hat{\mathbf{x}}_{k|k}}{\partial \hat{\mathbf{x}}_{k-N|k}} \end{pmatrix}}^{n} \left.\vphantom{\begin{pmatrix}a\\a\\a\\a\\a\end{pmatrix}}\right\} N+1 \tag{35}$$

und

$$\frac{\partial\,\hat{\mathbf{v}}}{\partial\,\hat{\mathbf{a}}} = \overbrace{\begin{pmatrix} \left(-\frac{\partial\,h(\hat{\mathbf{x}}_{k-N|k})}{\partial\,\hat{\mathbf{x}}}\right)^{\mathrm{T}} \frac{\partial\,\hat{\mathbf{x}}_{k-N|k}}{\partial\,\hat{\mathbf{a}}} \\ \left(-\frac{\partial\,h(\hat{\mathbf{x}}_{k-N+1|k})}{\partial\,\hat{\mathbf{x}}}\right)^{\mathrm{T}} \frac{\partial\,\hat{\mathbf{x}}_{k-N+1|k}}{\partial\,\hat{\mathbf{a}}} \\ \vdots \\ \left(-\frac{\partial\,h(\hat{\mathbf{x}}_{k|k})}{\partial\,\hat{\mathbf{x}}}\right)^{\mathrm{T}} \frac{\partial\,\hat{\mathbf{x}}_{k|k}}{\partial\,\hat{\mathbf{a}}} \end{pmatrix}}^{r} \Bigg\} N+1. \tag{36}$$

Dabei ist (35) wie beim *GNMHO* von der Lösung des matrixwertigen Differentialgleichungssystems abhängig:

$$\begin{aligned} \dot{\mathbf{X}} &= \frac{\partial\,\mathbf{f}(\hat{\mathbf{x}})}{\partial\,\hat{\mathbf{x}}}\mathbf{X}, \\ \mathbf{X}(0) &= \mathbf{I}. \end{aligned} \tag{37}$$

Die $(N+1, r)$-Matrix (36), die die Ableitung der Schätzfehler nach den Schätzungen der Parametervektoren enthält, wird durch die Lösung des matrixwertigen Differentialgleichungssystems:

$$\begin{aligned} \dot{\mathbf{Y}} &= \frac{\partial\,\mathbf{f}(\hat{\mathbf{x}})}{\partial\,\hat{\mathbf{x}}}\mathbf{Y} + \frac{\partial\,\mathbf{f}(\hat{\mathbf{x}})}{\partial\,\hat{\mathbf{a}}}, \\ \mathbf{Y}(0) &= \mathbf{0} \end{aligned} \tag{38}$$

bestimmt.
Bei einem Beobachter mit Schätzung der Parameter kommt man auf $n+n^2+nr$ zu simulierende Differentialgleichungen. Vergleicht man diese Summe mit dem Rechenaufwand der Parameterschätzung durch Zustandserweiterung, so erkennt man, dass der Aufwand hier wesentlich geringer ist. Ein weiterer Vorteil ist die Zahl der Messwerte, die für die Beobachtbarkeit notwendig ist. Da sie jeweils von der Systemordnung abhängig ist, ist sie für das zuerst beschriebenen Verfahren größer.

4 Erprobung der Verfahren

Die Erprobung der Verfahren wurden am Versuchsaufbau Dreitanksystem des Instituts für Automatisierungstechnik durchgeführt. Das mathematische Modell des Dreitanksystems ist in Gleichung (39) angegeben:

$$\begin{aligned} \dot{x_1} &= -V_1\sqrt{x_1+h_1} - \mathrm{sign}(x_1-x_2)\left(-\frac{a_{12}}{2b_{12}} + \sqrt{\left(\frac{a_{12}}{2b_{12}}\right)^2 - \frac{g|x_1-x_2|}{a_{12}}}\right) + p_1u_1, \\ \dot{x_2} &= -V_2\sqrt{x_2+h_2} + \mathrm{sign}(x_1-x_2)\left(+\frac{a_{12}}{2b_{12}} + \sqrt{\left(\frac{a_{12}}{2b_{12}}\right)^2 - \frac{g|x_1-x_2|}{a_{12}}}\right) \\ &\quad -\mathrm{sign}(x_2-x_3)\left(-\frac{a_{23}}{2b_{23}} + \sqrt{\left(\frac{a_{23}}{2b_{23}}\right)^2 - \frac{g|x_2-x_3|}{a_{23}}}\right), \\ \dot{x_3} &= -V_3\sqrt{x_3+h_3} + \mathrm{sign}(x_2-x_3)\left(-\frac{a_{23}}{2b_{23}} + \sqrt{\left(\frac{a_{23}}{2b_{23}}\right)^2 - \frac{g|x_2-x_3|}{a_{23}}}\right) + p_2u_2. \end{aligned} \tag{39}$$

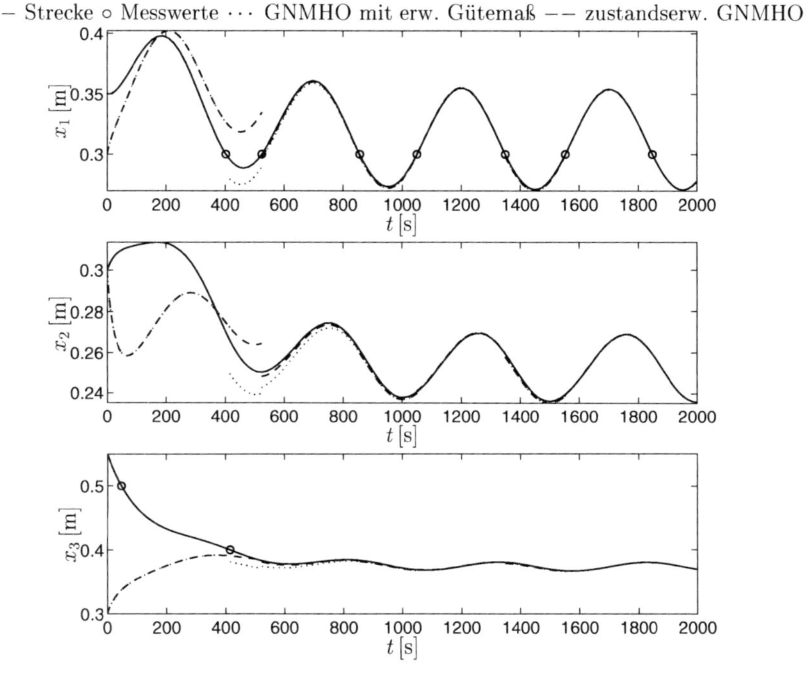

Abbildung 1: Vergleich der Parameterschätzverfahren anhand einer Simulation

Zunächst soll die Wirksamkeit beider Verfahren mit Hilfe einer Simulation nachgewiesen werden. Dabei wird der Parameter V_1 aus (39) als unbekannt angenommen. Er soll bei der Zustandsbeobachtung mitgeschätzt werden.

Für das Verfahren nach Abschnitt 2.3 mit der Systemordnung $n + r = 3 + 1 = 4$ wurde dabei der minimale Beobachtungshorizont von $N + 1 = n + r = 4$ festgelegt, für das Verfahren nach 3 mit der Systemordnung $n = 3$ reicht dagegen ein Beobachtungshorizont von $N + 1 = n = 3$ aus. Pro Beobachtungshorizont wurden jeweils zwei Gauss-Newton-Schritte ausgeführt. In Abbildung 1 sieht man wie beide Verfahren konvergieren. Der auf dem *GNMHO* mit erweitertem Gütemaß basierende Ansatz verbessert den Schätzwert schon nach drei Messwerten, während bei der Parameterschätzung durch Zustandserweiterung erst nach vier Messwerten ein verbesserter Schätzwert berechnet wird. Der rechnerische Aufwand beträgt für den zustandserweiterten *GNMHO* $n+r+(n+r)^2 = 3+1+(3+1)^2 = 20$ zu lösende Differentialgleichungen, wogegen für den *GNMHO* mit erweitertem Gütemaß $n + n^2 + n \cdot r = 3 + (3)^2 + 3 \cdot 1 = 15$ Differentialgleichungen simuliert werden müssen. Zusätzlich ist hier der Beobachtungshorizont kürzer als beim zustandserweiterten *GNMHO*.

Neben der Simulation der Parameterschätzung wurden beide Verfahren am Versuchsaufbau Dreitanksystem erprobt. Dabei wurde nicht nur der Parameter V_1 sondern auch V_2 (39) geschätzt. Pro Beobachtungshorizont wurde diesmal aber nur ein Gauss-Newton-

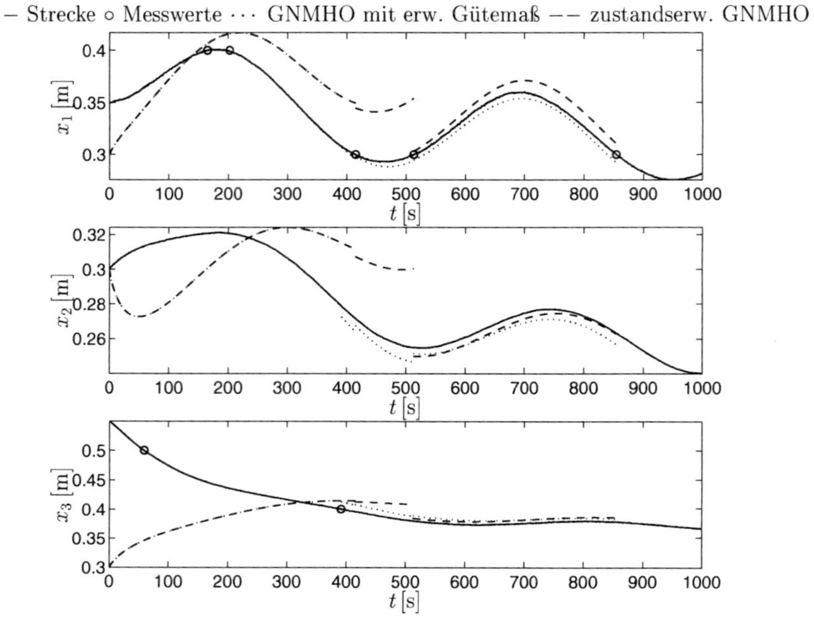

Abbildung 2: Vergleich der Parameterschätzverfahren am Versuchsaufbau Dreitanksystem

Schritt durchgeführt. Wegen der Systemordnung von $n + r = 3 + 2 = 5$ ist der minimale Beobachtungshorizont für den zustandserweiterten *GNMHO* in diesem Fall $N + 1 = 5$. Wegen der besseren Konvergenz wurde für den *GNMHO* mit erweitertem Gütemaß ein Beobachtungshorizont von $N + 1 = 4$ gewählt. Die sich für die Zustandsschätzung ergebende unterschiedlich schnelle Konvergenz kann man in Abbildung 2 sehen. Der rechnerische Aufwand beträgt für den zustandserweiterten *GNMHO* $n + r + (n + r)^2 = 3 + 2 + (3 + 2)^2 = 30$ zu lösende Differentialgleichungen, wogegen für den auf dem *GNMHO* mit erweitertem Gütemaß basierenden Ansatz $n + n^2 + n \cdot r = 3 + (3)^2 + 3 \cdot 2 = 18$ Differentialgleichungen über einen kürzeren Beobachtungshorizont simuliert werden müssen.

5 Zusammenfassung

Es wurde gezeigt, wie mit dem *GNMHO* zwei Methoden zur gleichzeitigen Zustands- und Parameterschätzung zur Verfügung stehen. Bei der Anwendung der Verfahren zeigt sich das auf dem *GNMHO* aufbauende Verfahren mit erweitertem Gütemaß (Abschnitt 3) in allen Punkten überlegen. Zum einen ist der Rechenaufwand niedriger zum anderen besitzt das Verfahren ein, verglichen mit dem zustandserweiterten *GNMHO* (Abschnitt 2), besseres Konvergenzverhalten (siehe Tabelle 1).

Tabelle 1: Vergleich der Parameterschätzverfahren

Parameter-schätz-verfahren	Zahl der Differential-gleichungen	Beobachtungs-horizont $N+1$	Ergebnisse (siehe Beispiel Dreitanksystem)
zustandserweiterter GNMHO	$n+r+(n+r)^2$	mind. $n+r$	relativ langsame Konvergenz
GNMHO mit erweiter-tem Gütemaß	$n+nr+n^2$	mind. n	relativ schnelle Konvergenz

Literatur

[1] C. Bohn. *Recursive Parameter Estimation for Nonlinear Continuous-Time Systems through Sensitivity-Model-Based Adaptive Filters.* 2000. Dissertationsschrift Universität Bochum.

[2] M.A. Henson (Ed.), D.E. Seborg (Ed.), K.R. Muske, and T.F. Edgar. *Nonlinear process control,* chapter Nonlinear state estimation, pages 311–370. Prentice-Hall, London, 1996.

[3] R. Isermann. *Identifikation dynamischer Systeme 1.* Springer, Berlin, 1992.

[4] R. Isermann. *Identifikation dynamischer Systeme 2.* Springer, Berlin, 1992.

[5] T. Krantz. *Entwicklung und Vergleich von Verfahren zur In-Orbit-Bestimmung des magnetischen Dipolmoments eines Satelliten.* Institut für Automatisierungstechnik, 2002. Diplomarbeit Universität Bremen.

[6] K.R. Muske and J.B. Rawlings. *NATO ASI Series,* volume 293 of *Series E,* chapter Nonlinear Moving Horizon State Estimation, pages 349–365. Kluwer Academic Publishers, 1995.

[7] J. Vortisch and B. Lohmann. A Gauss-Newton Moving Horizon Estimator and reduced order realizations in comparison to EKF. *Proceedings of the IEEE international Conference on Methods and Models in Automation and Robotics, Szczecin, Poland,* pages 1251–1256, 2002.

[8] G. Zimmer. *Zustandsbeobachtung nichtlinearer Systeme.* Shaker Verlag, Aachen, 1997. Habilitationsschrift TU Berlin.

Model based adaptive calculation of Lyapunov exponents from chaotic time series

M. Ataei
Institute of Automation, University of Bremen
Otto-Hahn-Allee. / NW1, D-28359, Bremen, Germany
(K. N. Toosi University of Technology, Sayyed Khandan Bridge, P.O. Box 16315-1355, Tehran, Iran)
E-mail: ataei@iat.uni-bremen.de

Abstract

The Lyapunov exponents (LEs) are the most basic dynamical diagnostic for deterministic chaotic systems. In this paper, an adaptive method for LEs computation from output time series based on a Jacobian approach is presented. This is achieved by properly formulating the regression vector of a nonlinear model and adaptively updating the parameters vector using the Recursive Least Squares (RLS) principles. Then, by a state space realization in canonical form, the Jacobian matrices used in a QR factorization method to calculate the LEs are computed. Finally, simulation results are presented for a well-known chaotic system with time varying parameters to show the effectiveness of the proposed adaptive methodology.

Key words: Chaotic time series, Lyapunov exponents, Jacobian matrices, QR factorization, Recursive Least Squares (RLS), nonlinear regressive

1 Introduction

The Lyapunov exponents are conceptually the most basic and useful dynamical diagnostic for deterministic chaotic systems. The calculation of LEs for systems whose dynamical equations with constant parameters are known is straightforward. However, these methods can not be applied directly to measurement data. Two general approaches for computing the LEs from output time series are the *geometrical* and the *Jacobian approaches.* In the geometrical approach, the long term evolution of an infinitesimal sphere of initial conditions is considered [Wolf *et al.,* 1985], [Rosenstein *et al.,* 1993], [Kantz, 1994]. However, in the Jacobian approach, local Jacobian matrices are estimated and the long term product of matrices is computed [Eckmann *et al.,* 1986], [Brown *et al.,* 1991], [Darbyshire, 1994], [Oiwa & Fiedler-Ferrara, 1998].

The procedure adopted in this paper falls into the Jacobian approach category. It is shown that this approach has the capability of adaptively computing the LEs of uncertain or time varying chaotic dynamical systems. The important step in this approach is to estimate the Jacobian matrices. Since the LEs are derived from the eigenvalues of the Jacobians, any small error in the computation of Jacobians can cause a major error in the LEs computation. Some general perturbation results and error analysis in QR algorithms for computing LEs can be found in [Dieci & Van Vleck, 1995], [Dieci *et al.,* 1997].

In the previous work on this subject, it is usually tried to find the Jacobians by *locally linear mapping* of neighbourhoods near the reference trajectory to neighbourhood at a subsequent time [Bryant *et al.,* 1990]. In [Sano & Sawada, 1985] and [Eckmann *et al.,* 1986], the linearized flow map from the neighbour data set into m step ahead of this set is considered as an approximation for the tangent map. In [Brown *et al.,* 1991], it is shown to be advantageous to use local neighbourhood-to-neighbourhood mappings with higher order Taylor series, rather than just local linear maps. But, these methods involve some subjects like state space reconstruction and finding the proper neighbours which have underlying problems. To have a proper reconstructed state space, the determination of embedding dimension and lag time have important roles. There are many methods that concern this topic including the *False Neighbour* method [Kennel *et al.,* 1992] and the *Singular Value Decomposition* method [Broomhead & King, 1986], [Mess *et al.,* 1987]. Looking for the suitable neighbourhoods also needs high computational efforts as well as time which is not unavoidable in above mentioned methods. Hence, having a method like an adaptive reconstruction of chaotic attractors from a single trajectory presented in [Zoltowski, 2000] can resolve these problems. Four other methods to estimate the Jacobian have been referred to in [McCaffrey, 1992], including the local thin-plate splines, radial basis functions, projection pursuit and neural nets. In [Oiwa & Fiedler-Ferrara, 1998], the Jacobians are estimated over boxes of the state space to speed up the algorithm of LEs computation.

However, in all the previous work it is generally assumed that the dynamical system being studied has fixed parameters and is time invariant. In this paper, the calculation of LEs is based on adaptive estimated Jacobian matrices. In fact, the goal is that by using a known non-linear structure for the chaotic dynamical equations and recursively estimating the unknown parameters of the model, the procedure of the Jacobian estimation is performed on-line to overcome the problem of time varying and unknown plant parameter problems. Hence, the calculation of LEs based on these Jacobians and by using QR factorizaton is done adaptively. Therefore, any variations in unknown physical parameters of the system will appear on-line on the LEs, i.e, the underlying LEs of the system for the current parameters are available and can be used for predictability analysis.

The background materials are given in Sec. 2. An adaptive algorithm for the calculation of the LEs is presented in Sec. 3. Finally, simulation results are provided to show the effectiveness of the proposed methodology in Sec. 4.

2 Background Materials

Consider the autonomous discrete dynamical system described in the following form:

$$X_{k+1} = F(X_k) \quad k = 0,1,\cdots \tag{1}$$

where X_k is the state vector in the R^m space and $F(\cdot)$ is a continuously differentiable nonlinear function. Linearization of the system for a small range around the operational trajectory in the phase space can be written as:

$$X_{k+1} = J_k \cdot X_k \quad k = 0,1,\cdots \tag{2}$$

where $J_k = \left.\frac{\partial F}{\partial X}\right|_{X_k} \in R^{m\times m}$ is the Jacobian matrix in point k. The LE is defined as follows [Dieci & Van Vleck, 1995]:

Definition 1- Let $Y^k = J_{k-1}\cdot J_{k-2}\cdots J_0$, then the following symmetric positive definite matrix exists:

$$\Lambda = \lim_{k\to\infty}\left(\left(Y^k\right)^T\cdot Y^k\right)^{\frac{1}{2k}} \tag{3}$$

and the logarithms of its eigenvalues are called the *Lyapunov Exponents*.

However, there are some problems with the computation of the LEs by using this definition. The first problem is that for a large value of k, the fundamental solution Y^k may result in very large values and actually the calculation of Λ with this format is not possible. Further, the computation of Y^k should be such that the linear independency of the columns is maintained. Otherwise, this computation leads only to the largest LEs. To deal with these problems, the QR factorization algorithm is used for approximation of LEs [Eckmann *et al.,* 1986], [Brown *et al.,* 1991], [Darbyshire, 1994], [Oiwa & Fiedler-Ferrara, 1998], [Dieci & Van Vleck, 1995], [Dieci *et al.,* 1997]. The steps involved in this method can be summarized as follows [Dieci & Van Vleck, 1995]:

1- Given orthogonal Q_0 such that $Q_0^T\cdot Q_0 = I$.

2- Solve $Z_{k+1} = J_k\cdot Q_k$, $k = 0,1,\cdots$ and obtain the decomposition: $Z_{k+1} = Q_{k+1}\cdot R_{k+1}$ where Q_k is an orthogonal matrix and R_{k+1} is an upper triangular matrix with positive diagonal elements.

3- $\lambda_i = \lim_{k\to\infty}\frac{1}{k}\log((R_k)_{ii}\cdots(R_1)_{ii}) = \lim_{k\to\infty}\frac{1}{k}\sum_{j=1}^{k}\log((R_j)_{ii})\quad i = 1,\cdots,m$

3 Adaptive Calculation of LEs

This section presents the adaptive calculation of the LEs. It is supposed that the output data of the dynamical system is available as a univariate time series. The dynamical behavior of the system is described with the following nonlinear difference equation:

$$y(k+1) = f(Y(k)) \tag{4}$$

where $f(\cdot)$ is a continuously differentiable function and $Y(k)$ is a delayed vector as:

$$Y(k) = [y(k-m+1), y(k-m+2), \cdots y(k)] \tag{5}$$

The value of m can be selected as the minimum dimension of the embedding space to have a reconstruction of the original state space. In practice, usually the structure of the underlying dynamical system which generates the data is not known. Depending on the objectives, there are different theories, such as the functional theory, which are suitable for special analysis of nonlinear systems. By this theory, the dynamical behavior of the nonlinear processes can be described by the Voltera functional series. Then, the nonlinear process model with a finite number of parameters and linear in the unknown parameters can be derived from the discrete Voltera series [Isermann *et al.,* 1992]. Here, a definite structure for the nonlinear autoregressive model (4) is assumed as follows:

$$\begin{aligned} y(k+1) &= \sum_{i=0}^{m-1}\theta_{1i}y(k-i)+\sum_{\beta=0}^{h}\sum_{i=0}^{m-1}\theta_{2\beta i}y(k-i)y(k-i-\beta)+\cdots \\ &+\sum_{\beta_1=0}^{h}\sum_{\beta_2=\beta_1}^{h}\cdots\sum_{\beta_{p-1}=\beta_{p-2}}^{h}\sum_{i=0}^{m-1}\theta_{p\beta_1\ldots\beta_{p-1}i}y(k-i)y(k-1-\beta_1)\cdots y(k-1-\beta_{p-1})+\theta_0 \end{aligned} \tag{6}$$

By considering $Y(k)$ in (5) as the state vector, a canonical state space representation of the system is obtained as follows:

$$X(k)=\begin{bmatrix} x_1(k) \\ x_2(k) \\ \vdots \\ x_m(k) \end{bmatrix}=\begin{bmatrix} y(k-m+1) \\ y(k-m+2) \\ \vdots \\ y(k) \end{bmatrix} \Rightarrow X(k+1)=\begin{bmatrix} x_2(k) \\ x_3(k) \\ \vdots \\ f(X(k)) \end{bmatrix} \tag{7}$$

The Jacobian matrix J_k in each point k of the typical trajectory for this canonical representation is as:

$$J_k=\begin{bmatrix} 0 & 1 & \cdots & 0 & 0 \\ 0 & 0 & 1 & \cdots & 0 \\ \vdots & \vdots & \vdots & \ddots & \vdots \\ Df_1 & Df_2 & \cdots & Df_{m-1} & Df_m \end{bmatrix} \quad \text{where } Df_i=\frac{\partial f}{\partial x_i}. \tag{8}$$

Assuming the structure given by Eq. (6), the $Df_i \quad i=1,\ldots,m$ are known parameters in terms of the parameters of the model. Therefore, to have the Jacobians in each point of the trajectory, only the recursive estimation of unknown parameters in (6) is required. To achieve this, the RLS algorithm is used. By defining the regression vector as:

$$\phi=\left(y(k),\cdots,\, y(k-m+1),\, y^2(k),\cdots,\, y^2(k-m+1),\cdots,\, y(k-m+1)y(k-h-m+1),\cdots\right)^T \tag{9}$$

and considering the underlying parameter vector θ, the RLS method is used to estimate the vector θ as follows [Landau *et al.,* 1998]:

$$\hat{\theta}(k+1)=\hat{\theta}(k)+F(k+1)\phi(k)\varepsilon^0(k+1) \tag{10}$$

where:

$$F(k+1)=F(k)-\frac{F(k)\phi(k)\phi^T(k)F(k)}{1+\phi^T(k)F(k)\phi(k)}$$
$$\varepsilon^0(k+1)=y(k+1)-\hat{\theta}^T(k)\phi(k) \tag{11}$$

Now, by using the estimated parameters, the QR method for calculation of the LEs can be modified as an *adaptive algorithm* for the computation of the LEs as follows:

Algorithm 1:

1- In step k by using the relation (10), the unknown parameters are estimated. Therefore, the function $f(\cdot)$ according to the difference equation (6) is known.

2- The Jacobian matrix J_k is computed and the decomposition $J_k \cdot Q_k = Q_{k+1} \cdot R_{k+1}$ is obtained where Q_k is an orthogonal matrix and R_{k+1} is a upper triangular matrix with positive diagonal elements.

3- The LEs are calculated adaptively:

$$\lambda_i(k+1)=\frac{1}{k+1}\left(k\cdot\lambda_i(k)+\log\left(R(k+1)_{ii}\right)\right) \quad i=1,\cdots,m \tag{12}$$

In the face of system parameter variations, the LEs of the system will change and the proposed adaptive algorithm shall identify the new LE. Therefore, calculating the LEs adaptively makes it possible to have the true value in each step of time and this facilitates an accurate predictability analysis of the data generator chaotic system with time varying parameters. Also, computing the Jacobians do not rely on finding the local map between neighbours of the reference point and the neighbours in subsequent times. This requires much computational efforts and takes a lot of time that has been avoided in the adaptive method.

4 Simulation Results

To show the effectiveness of the proposed adaptive calculation of LEs, the algorithms are applied to a well-known chaotic system. The Henon map is considered with following difference output equation:

$$y(k+1)=1-ay^2(k)+by(k-1) \tag{13}$$

where a and $|b|\le 1$ are unknown parameters which are time varying. Suppose that the nominal parameters are $a=1.4, b=0.3$ which after some steps change to new values as $a=1, b=0.1$. Fig. 1 shows the graph of the output data around the region where the nominal parameters have been changed to their new values. In practical systems, this kind of changes in nominal parameters is a common phenomenon which occurs due to time dependent variations in the physical quantities of the system. This causes variations in the LEs of the

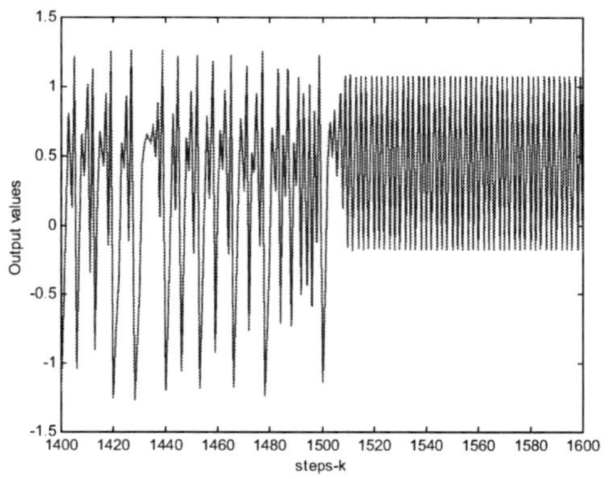

Fig.1. Data of Henon map with changes in parameters in step $k = 1500$.

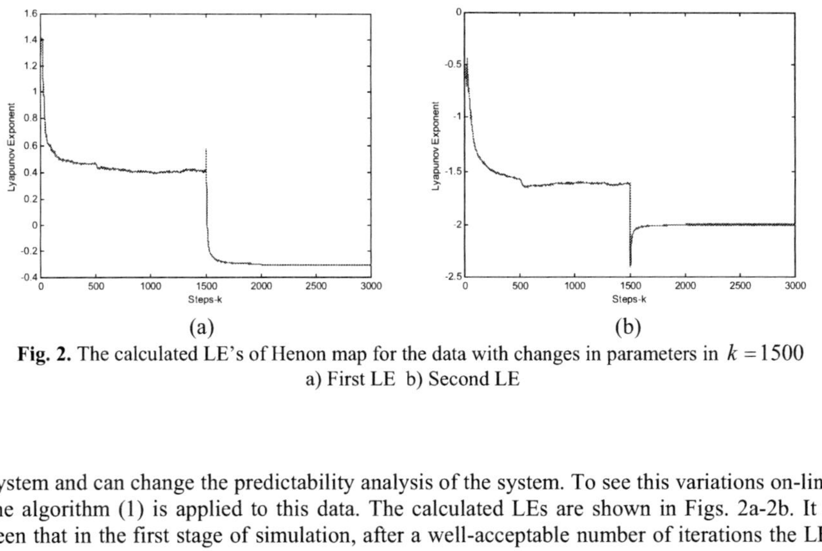

Fig. 2. The calculated LE's of Henon map for the data with changes in parameters in $k = 1500$
a) First LE b) Second LE

system and can change the predictability analysis of the system. To see this variations on-line, the algorithm (1) is applied to this data. The calculated LEs are shown in Figs. 2a-2b. It is seen that in the first stage of simulation, after a well-acceptable number of iterations the LEs have converged to the exact values which are $\lambda_1 = 0.42, \lambda_2 = -1.62$.

In $k = 1500$, the external parameters change to the new values $a = 1, b = 0.1$. The calculated LEs are changed iteratively to converge to the correct LEs for these parameters. It is shown that this convergence is very fast and the exact values of LEs, $\lambda_1 = -0.306, \lambda_2 = -1.99$, for this case is achieved in $k = 2000$. It is seen that in this case both LEs are negative and the system has exited from the chaotic range.

5 Summary

In this paper, an adaptive approach for the calculation of LEs, is proposed. This ensures the effective calculation of LEs in the face of system parameter variations. The adaptive methodology is based on a non linear regression vector and the RLS algorithm for the on-line parameter update. Finally, to show the effectiveness of the proposed adaptive methodology, it is applied to the well-known Henon chaotic system and the simulation results are provided to present the main points of the paper.

References

Broomhead, D. S., & King, G. P. [1986] ''Extracting qualitative dynamics from experimental data'', *Physica* ***D*** **20**, 217-236.

Brown, R., Bryant, P., & Abarbanel, H. D. I. [1991] ''Computing the Lyapunov spectrum of a dynamical system from an observed time series'', *Phys. Rev.* ***A*** **43**(6), 2787-2806.

Bryant, P., Brown, R., & Abarbanel, H. D. I. [1990] ''Lyapunov exponents from observed time series'', *Phys. Rev. Let.* **65**(13), 1523-1526.

Darbyshire, A. G. [1994] ''Calculating Liapunov exponents from a time series'', *IEE Digest, No,* 1944/13, 5.

Dieci, L., Russell, R. D., & Van Vleck, E. S. [1997] ''On the computation of Lyapunov exponents from continuous dynamical systems'', SIAM J. Numer. Anal., **34**(1),402-423.

Dieci, L., & Van Vleck, E. S. [1995] ''Computation of few Lyaounov exponents for continous and discrete dynamical systems'', *Appl. Numer. Math.* **17**, 275-291.

Eckmann, J. P., Kamphorst, S. O., Ruelle, D., & Ciliberto, S. [1986] ' Liapunov exponents from time series' *Phys. Rev.* ***A*** **34**(6), 4971-4979.

Isermann, R., Lachmann, K. H., & Matko, D. [1992] *Adaptive Control Systems*, Prentice Hall, Chap. 17.

Kantz, H. [1994] ''A robust method to estimate the maximum Lyapunov exponent of a time series'', *Phys. Lett.* ***A*** **185**, 77-87.

Kennel, M. B., Brown, R., & Abarbanel, H. D. I [1992] ''Determining embedding dimension for phase space reconstruction using a geometrical construction'', *Phys. Rev.* ***A*** **45**(6), 3403-3411.

Landau, I. D., Lozano, R., & M'Saad, M. [1998] *Adaptive Control*, Springer-Verlag, Chap. 3.

McCaffrey, D. F., Ellner, S., Gallant, A. R., & Nychka, D. W. [1992] ''Estimating the Lyapunov exponent of a chaotic system with nonparametric regression'', *Journal of the American Statistical Association*, **87**(419), 682-695.

Mees, A. I., Rapp, P. E., & Jenning, L. S. [1987] ''Singular value decomposition and embedding dimension'', *Physical review A*, **36**(1), 340-346.

Oiwa, N. N., & Fiedler-Ferrara, N. [1998] ''A fast algorithm for estimating Lyapunov exponents from time series'', *Phys. Lett.* ***A*** **246**, 117-121.

Rosenstein, M. T., Collins, J. J., & De Luca, C. J. [1993] ''A practical method for calculating largest exponents from small data set'', *Physica* ***D*** **65**, 117-134.

Sano, M., & Sawada, Y. [1985] ''Measurement of the Lyapunov spectrum from a chaotic time series'', *Phys. Rev. Let.* **55**(10), 1082-1085.

Wolf, A., Swift, J. B., Swinney, H. L., & Vastano, J. A. [1985] ''Determining Lyapunov exponents from a time series'', *Physica* ***D*** **16**, 285-317.

Zoltowski, M. [2000], ''An adaptive reconstruction of chaotic attractors out of their single trajectories'', *Signal Processing*, Vol. 80, 1099-1113.

Soft Computing Techniques are Essential in Human Centered Human-Robot Interaction

Z. Zenn Bien, Kwang-Hyun Park, Dae-Jin Kim, Jin-Woo Jung, Jeong-Su Han
Jun-Hyong Do and Hyoyoung Jang
Division of Electrical Engineering,
Department of Electrical Engineering and Computer Science
Korea Advanced Institute of Science and Technology (KAIST),
373-1 Guseong-dong, Yuseong-gu, Daejeon, 305-701, Republic of Korea,
zbien@ee.kaist.ac.kr, akaii@robotian.net
and {djkim, jinwoo, pyodori, jhdo, hyjang}@ctrsys.kaist.ac.kr

Friendly interaction between human beings and robotic systems is vital in designing human centered service robots. Among several interaction technologies, intention reading of the user plays an important role for the human centered systems. In this paper, we focus on the aspect of intention reading in service robotic systems, and soft computing techniques are adopted to deal with such a system. An effective intention reading scheme is proposed on the basis of several soft computing techniques to handle uncertainty of the user's intention and is applied to several service robotic systems.

1 Introduction

Lately, the focus of advanced research activities in designing automation systems has been shifted from production and manufacturing areas toward service sectors. One of the popular service automation systems under particular attention is a service robot system. A service robot is a re-programmable, sensor-based mechatronic system that can perform useful works to human activities [1]. Since functions of service robots are generally related to the ordinary human life such as repair, transfer, cleaning, and health care, human beings and robots may coexist and collaborate in the same environment, and close interaction may be supposed to take place. Due to variability of the user's physical motion and uncertainty of the environment, it is usually difficult to model and handle interaction between human beings and robots.

To resolve such a complexity in designing human-robot interaction systems, one powerful approach is to view human beings and robots as a human-in-the-loop system. Most conventional human-in-the-loop systems are machine centered systems as shown in Fig. 1, and the human being is considered as a bad component due to time-varying, inconsistent and fuzzy characteristics in decision-making. One of the primary objectives of such conventional systems is to reduce the human factors as much as possible so as to minimize any possible errors and unrelated effects. Thus, human beings should learn complicated procedures of all the machines to be precise and time-invariant. An airplane control system can be a typical example of machine sensor type systems and an overhead crane system [2] can be an exemplary human sensor type system. In these systems, machine sensors are accurate, but not any more effective in complicated environment due to limitation of intelligence and in processing information. Human sensors are efficient in getting high-level information, but takes cognitive load during the operation.

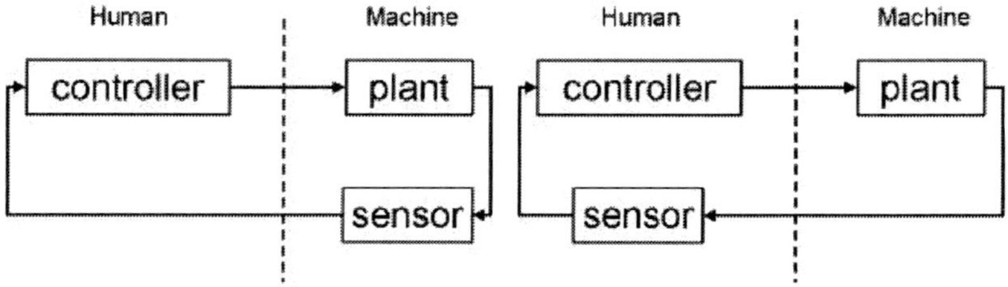

(a) Machine Sensor Type (b) Human Sensor Type

Fig. 1 Machine Centered System

Human centered system, shown in Fig. 2, can be considered as an alternative human-in-the-loop system whose components are designed and implemented in a human-friendly manner. Unlike conventional human-in-the-loop systems, the human centered system can reduce the user's cognitive load during the operation of the overall system and the machine should be designed to have capabilities of convenience, safety and comfort. In order to be human-friendly, it is necessary to understand the human's intention and, in many cases, human's intention is represented by various kinds of biosigns such as a voice, gestures, facial expressions, eye-gazing, biosignals, and so on.

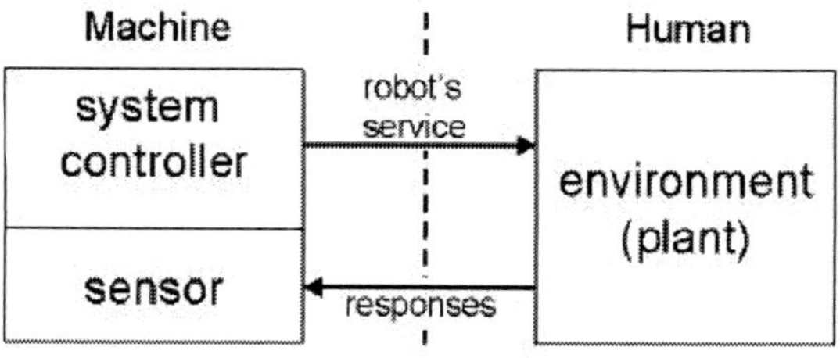

Fig. 2 Human Centered System

In this paper, we focus on the aspect of intention reading in the human centered service robotic systems, and soft computing techniques are adopted to deal with such a system. An effective intention reading scheme is proposed based on soft computing techniques to handle uncertainty of the user's intention and is applied to several service robotic systems for human-friendly interaction. This paper is organized as follows: In Section 2, intention reading between human beings is introduced. In Section 3, an effective intention reading scheme is proposed based on various soft computing techniques. In Section 4, several applications of intention reading are presented. Finally, concluding remarks are given in Section 5.

2 Understanding Human's Intention

A human being understands the other party's intention by interpreting various forms of information received during communication. Information exchange with the other person and interpretation of the received information are essential functions to understand human's intention. When a human being understands a partner's intention through communication, two loops of information processing may need to realize the above functions as shown in Fig. 3 [3]. First, human beings exchange information with each other through the information exchange loop. Information exchange is realized by means of language and behavior. Second, a human being understands the partner's intention in an intention understanding loop. Intention understanding is accomplished on the basis of knowledge concerning the partner.

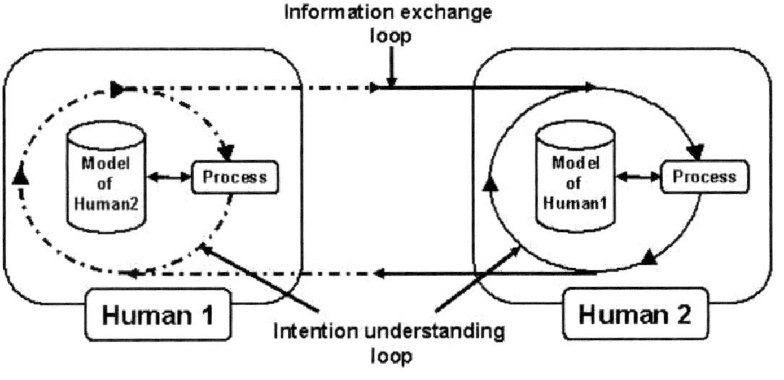

Fig. 3 Loops in communication between human beings [3]

In Fig. 3, suppose that the human1 has an intention and human2 reads the human1's intention through the intention reading process. In view of the human1 who has some intention, two behavioral patterns can appear to represent his/her intention; one is conscious behavior and the other is unconscious behavior [3]. The conscious behavior is a kind of consciously suppressed (or controlled) intentions. For example, a human operates various input devices such as a keyboard and a joystick in a conscious manner. On the contrary, the unconscious behavior is a kind of naturally happened intentions such as yawning when one feels sleepy.

In view of an observer, we may also say that human2 can observe two forms of intention according to how the intention is represented to the human2. One is an explicitly represented intention and the other is an implicitly represented intention [4]. An explicitly represented intention is clearly observable to the viewer. For example, a human issues commands such as "assist my walk" or "open the window" by voice. On the other hand, an implicitly represented intention is not clearly observable to the viewer, and there may involve much uncertainty. For example, if a person is stroking some place of his/her stomach with his/her hand, an agent can perform some operation on the conjecture that the human may have a stomachache.

According to the above classification, we may say that the intention reading typically refers to an ability to estimate an unconscious behavior or an implicitly represented intention. From an engineering point of view, intentions are conceptual, behavioral models for explaining, predicting and directing human actions within a defined context. Outstanding features of intentions are as follow [5]: First, intentions are not directly observable. Second, humans do not always verbalize their own intentions well. Third, intentions have a beginning and at least a hypothetical ending condition. Finally, individual differences have an influence on intention information. As humans do in the daily life, the partner's intention can be inferred not only from language but also from behavior. Usually, inferred intentions are vague and not necessarily correct, but it plays a key role in smooth cooperation for the user's comfort or in conservative decision-making for the user's safety. A human being also tries to read the other party's intention subjectively, and thus any classical mathematical approach may not be appropriate to express one's intention [6]. Hence, we need appropriate methods based on soft computing techniques in order to deal with these types of vague and uncertain knowledge.

3 Effective Human-Robot Interaction Scheme

In the human centered systems, intention reading can play a key role so as to change the human's vague and implicit expressions into formal commands to machines. Fig. 4 shows a model which describes signal flow from human's *mind level* to machine's *action decision making module*. Intention in mind level induces various biosigns through many human's physical organs such as face, hand, muscle, brain and vocal cord in the *body level*. These biosigns include facial expressions, eye-gazing, gestures, biosignals, and voice. The machine senses biosigns using various sensors in *acquisition module* and recognizes intention in the *intention reading module* (see Fig. 5). Finally, the machine's actions are made between human beings and service robots.

To deal with the biosign, which is complex, ambiguous, and susceptible, conventional mathematical approaches are not applicable and soft computing techniques are used in *intention reading module* and *action decision making module*. Soft computing techniques are convenient utensils to solve many real-world problems, and are in fact known to exploit the tolerance for uncertainty and imprecision to achieve tractability, robustness, and low solution cost [7]. Key methodologies include the fuzzy logic theory, neural networks, approximation reasoning, kinds of learning theory, derivative-free optimization methods such as evolutionary computation, and the rough set theory [8]. Complementary combination of these methodologies may show a strong computing power that "parallels the remarkable ability of the human mind to reason and learn in an environment of uncertainty of imprecision [9]."

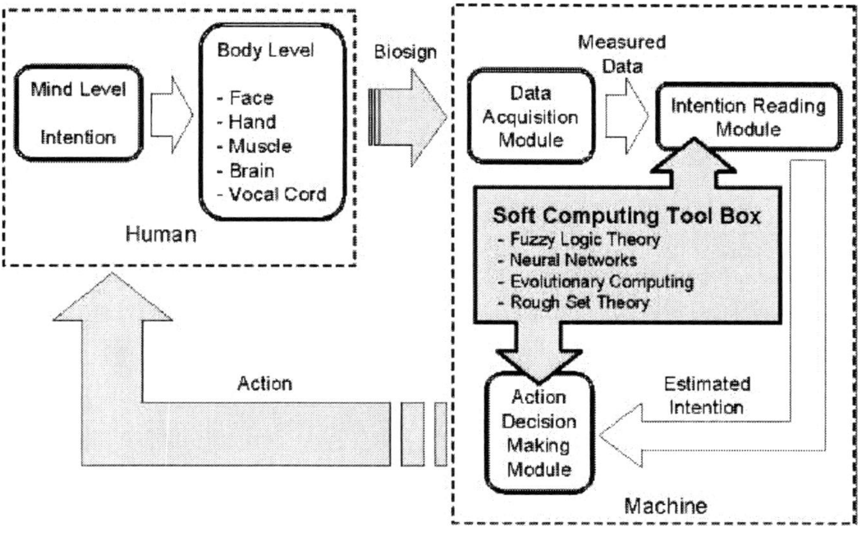

Fig. 4 Soft computing-based intention reading procedure from human mind level to action decision making level

For effective intention reading, we propose a systematic procedure in three steps: sensing, transformation, and recognition. First, the sensing step is for converting the physical data in the real world into measured data which will be used for further computation. The commonly used sensing device includes a CCD camera, a force/torque sensor, an infrared sensor, a strain gauge, pressure sensor, and so on. Second, those measured data are transformed into various kinds of features through the transformation step. Here, the data may be converted as abstract knowledge as human do, and we can use linguistic values for qualitative representation of abstract knowledge. Intention extraction is executed based on some soft computing techniques such as an artificial neural network or a fuzzy rule based system adopting a human expert's decision-making process. Then, we get a command to the machines by using crisp or fuzzy values. Fig. 5 shows the block diagram of our proposed intention reading scheme. Here, '^' denotes an estimated value. Thus, we can extract a user's intention from physically observable data through human-like decision-making system on the basis of soft computing techniques. The extracted intention can be a crisp number or a fuzzy number as human beings represent the other's intention from physical data with crisp or qualitative fuzzy values. Besides the artificial neural network and fuzzy logic theory, several soft computing techniques can be synergistically applied to implement this procedure, such as evolutionary computation, learning theory, and rough set theory.

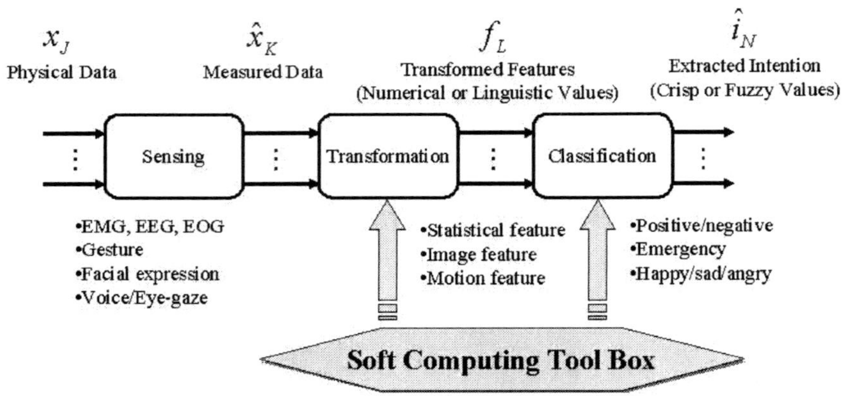

Fig. 5 Block diagram of proposed effective intention reading scheme

5 Applications to Service Robotic Systems

In this section, we describe several applications of intention reading for service robotic systems which are typical examples of human centered systems. In KARES (KAIST Rehabilitation Engineering Service system) II [10][11][12][13][14] and Intelligent Sweet Home [15], various intention reading systems are adopted to assist daily living activities and emotional satisfaction of the elderly and/or the handicapped person.

5.1 Facial Expression Recognition [16][17][18]

In general, the problem of recognizing emotion from a face is known to be very complex and difficult since individuality may come in expressing and observing emotions. It is, however, interesting to note that human beings can successfully understand facial expressions in a seemingly easy way. Thus, we develop a method to deal with such emotional information by observing the user's facial image and by using several well-known soft computing techniques [16][17].

As an practical application, we have been developing an intention reading system for intelligent visual servoing for wheelchair-mounted rehabilitation robot system, KARES II [18]. Among many principal tasks of KARES II, we focus on the task of "bringing the cup to the user's mouth". According to the extracted user's intention, the robot may approach to the user's face or get away from it. Detailed procedure is described in next subsections.

5.1.1 Sensing

As a sensing device, we use color CCD camera (CV-M2350 from JAI, co.) in eye-in-hand configuration on the robot's end-effector. The grabbed image is restricted to 640×480 (in pixel size) with a cluttered scene.

5.1.2 Transformation

From the grabbed images, we extract the user's intention by measuring user's facial aspect and the status of mouth. Fuzzy membership-based image features are used to measure the degrees of each facial aspect (i.e., left, center and right) [18]. To measure the mouth openness, a gabor-gaussian image feature is used [17]. Using gabor-gaussian image feature, we can easily know the status of mouth even though the individual difference of lips among many persons.

5.1.3 Classification

In classification stage, the user's facial aspect is determined by simple decision logic with the degrees of each facial aspect. To determine the degree of mouth openness, fuzzy rules are constructed by gathering many mouth images and extensive trials. Using 110 test images, we can successfully determine the user's mouth opening or closing with success ratio 92.1%.

5.1.4 Service Action

At last, according to the result of classification, the robot arm provides many kinds of service to the user. Here, we choose a 'serving a drink' as an example. The robot arm picks up a cup with beverage and moves it to the user's mouth. During this task, we can control the robot arm's act according to the user's facial aspect and the status of mouth.

Especially, service actions are composed as follows:

- According to the degree of mouth openness, robot arm approaches the user's mouth slowly or vice versa.
- When the mouth is almost closed, the robot arm doesn't move any more.

- If the user's facial aspect is constantly changing, i.e. shaking head, the robot arm retreats to its original position far away from the user's mouth.

As shown in Fig. 6, the robot arm acts successfully according to the user's intention.

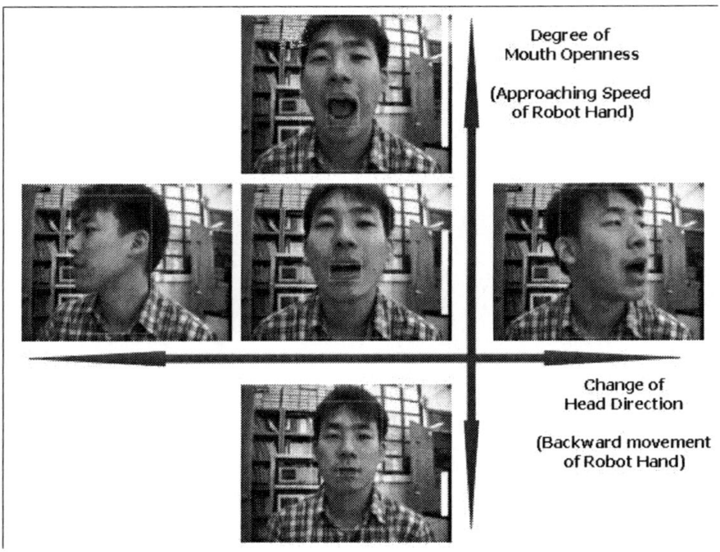

Fig. 6 Facial expression recognition: application to human-robot interaction based on user's intention reading

5.2 EMG-based Wheelchair/Robotic Arm Control [19][20][21]

EMG (electromyogram) signal is a form of electric manifestation of neuromuscular activation associated with contracting muscle [19]. The prostheses controller using EMG is a typical example that has been successfully developed to help the disabled in delivering his intention to nearby assistive device.

5.2.1 Sensing

To sense EMG signals, Ag/AgCl electrode and shielded cable are used to reduce the effect of noise in the environment. The locations of electrodes are pronator teres, brachialis, levator and deltoid, which have shown the highest contraction level for taking each predefined motion (see Fig. 7).

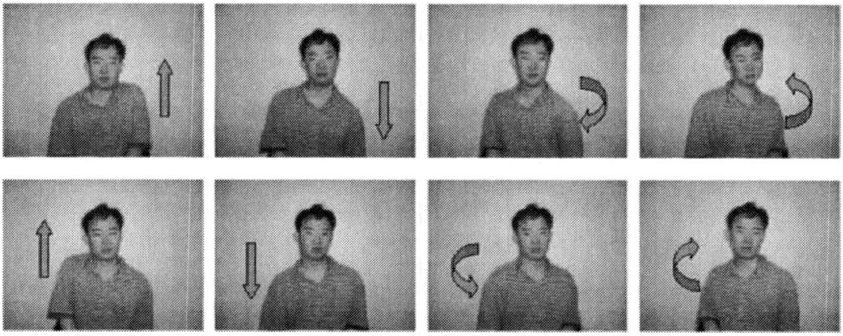

Fig. 7 Predefined eight basic motions

5.2.2 Transformation

To tackle the user dependency problem exhibited in many previous works [20], we propose an algorithm capable of classifying the biosignals obtained from different subjects into the predefined classes using a Fuzzy C-means algorithm and a rough set-based technique selecting a necessary and sufficient set of features out of all feature combinations extracted [20].

The overall signal processing procedure is briefly explained as follows. Well-known features such as integral absolute value (IAV), variance (VAR), zero crossing (ZC), frequency ratio (FR) are extracted to classify the predefined motions from measured EMG signals. By applying feature extraction algorithm [20] to these numerous extracted features, we can obtain the minimized feature combination sets which have enough information for complete classification (See Fig. 8).

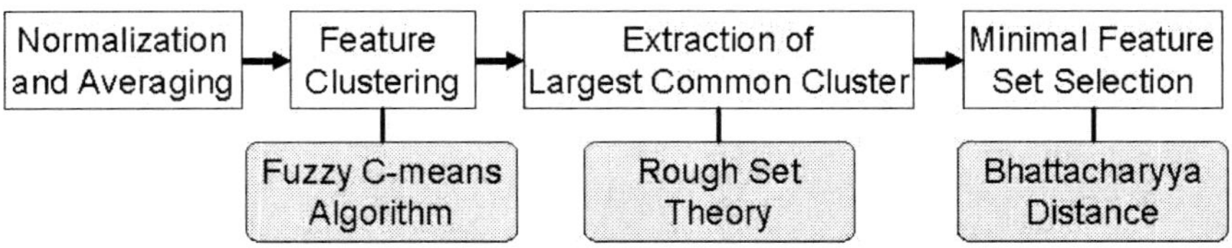

Fig. 8 The proposed algorithm of extracting feature set

5.2.3 Classification

The minimized feature combination sets extracted by the proposed algorithm are used as input-output pairs to make Fuzzy Min-Max Neural Networks (FMMNN) learn the motions.

5.2.4 Service Action

After learning, FMMNN gives the classified results and actuates the robotic arm or wheelchair [21] based on the user's movement. The basic motions are recognized with success rates of approximately 90% for four untrained subjects. We ascertained that EMG interface is useful for Human-Machine Interface (HMI) in KARES II.

5.3 Korean Sign Language Recognition System [22][23][24]

To overcome inconveniences of human-machine communication tools such as keyboards and mouse, the hand gesture method has been developed to accommodate a variety of commands naturally and directly. In spite of its usefulness, however, hand gesture is difficult to recognize by a machine due to its inconsistency, time-variance and so on [22].

Construction of a hand gesture recognition system involves structural categorization of gesture, real-time dynamic processing, pattern classification in a hyper dimensional space, coping with deterioration on recognition rate in case of expansion of gesture, dealing with ambiguity and nonlinearity constraints of the sensors, etc. Naturally, several intelligent processing methods such as soft computing technique have been evolved to overcome these difficulties [23][24]. Following subsections explain about the detailed procedure.

5.3.1 Sensing

A single color CCD camera (PULNIX) is used by capturing the image size of 600x400 (in pixel size). For fast image processing and image grabbing, Matrox Genesis^{TM} board is used through whole processes. Users don't need to wear special clothes and gloves at all. The background can be also allowed any cluttered scene.

5.3.2 Transformation

As a fundamental image feature, we extract silhouette images of both hands. From this, we acquire more detailed image features such as center-of-gravity (COG) of each hand, texture of hand surface, indirect information for the number of fingers and so on.

5.3.3 Classification

To classify the user's 18 hand-motion classes, HMM-based classifier is used with COG features. Here, HMM acts a key role to deal with such time-variant and unequally acquired sequential inputs. Average success ratio is 95.3%. Next, for 18 hand-pose classes, Fuzzy Min-Max Neural Networks (FMMNN)-based classifier is used. Texture of hand surface and other spatial features are inputted and average success ratio is 95.8%. At last, combining hand-motion and hand-pose, we can successfully classify 15 KSL sentences (94.7%) and 30 KSL words (96.4%).

5.3.4 Service Action

After classification, the KSL recognition system reports the result to the user in the form of voice signal and text in display (see Fig. 9). Also, the KSL recognition system can enable the deaf to communicate with the able-bodied person or vice versa. The deaf can understand the intention of the able-bodied person in remote site by 3D virtual character which can express various kinds of KSL words, sentences and facial expressions.

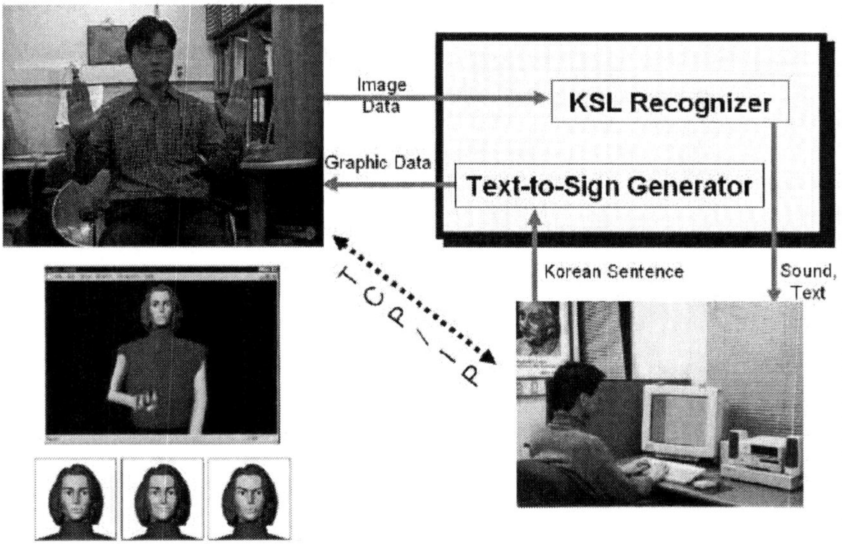

Fig. 9 Hand gesture-based communication between the deaf and the able-bodied

5.4 Hand Signal Recognition [25]

Hand-signal recognition is needed for a variety of applications and it is now gaining in importance as one method of enabling natural and intuitive human machine interface. User-interfacing method based on hand-signal recognition has more advantages than others, for example, commanding methods using voice, in the cases of:

- Emergencies
- Places where vocally communications are difficult such as a field of construction work
- Situations where the hand-signal is the only communication means, for example communication under water

The crucial point in recognizing hand-signals is that it requires great computational power and complicated structural analysis of the hand image. Our works deal with a new structure that is capable of recognizing hand-signals without any special glove or marker in 3-dimensional space under complex background and varying illumination.

5.4.1 Sensing

Two color CCD cameras are used to get the hand image sequence. Fig. 10 shows the configuration of those two cameras and captured images from each camera.

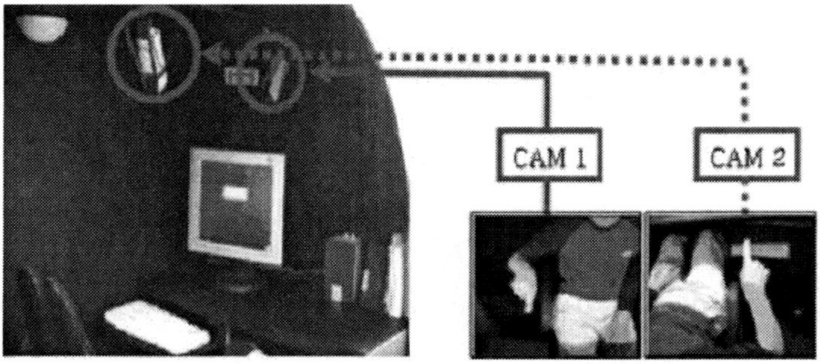

Fig. 10 Camera configuration and grabbed images

5.4.2 Transformation

In order to detect hand region robustly from an input image under complex background and varying illumination without any special glove or marker, we use previous hand position information and adaptive hand color model based on GMM (Gaussian Mixture Model). The center of gravity, elongate ratio, major axis, area and roughness of the segmented hand region are calculated as features of hand-signal.

5.4.2 Classification

To classify hand-signals, we define a hand-signal as a combination of two basic elements – 'hand pose' and 'hand trajectory'. We defined 32 hand poses and 5 hand trajectories. As an extensive classification method for hand pose, a multi-stage classification method by using small group concept has been developed. Also, a complementary selection method for two camera images has been used. We use HMM (Hidden Markov Model) in the hand trajectory classification.

5.4.3 Interpretation and Service Action

We verified the hand-signal system by applying to our driving simulator. With this system, it is possible to command two kinds of actions – device operation and communication to the other drivers outside the car. User can, for example, turn on the radio and control volume. Fig. 11 shows overall structure of the system.

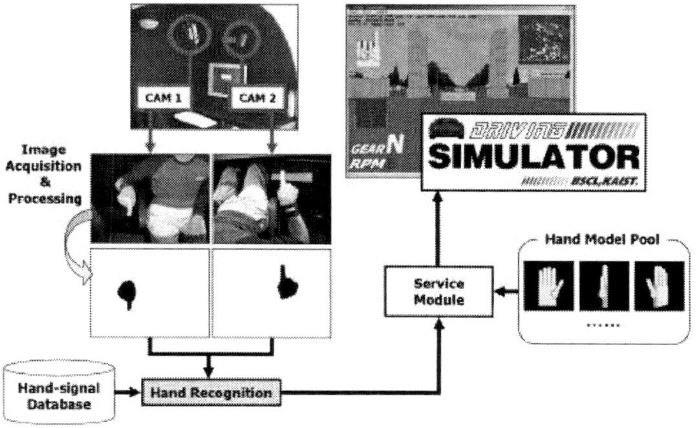

Fig. 11 Overall structure of the system

5.5 Emotional Interactive Robot [26][27][28]

Emotional Interactive Robot (EIR) is designed to understand the user's emotional condition and react depending on the user's emotional state. For emotion monitoring, it observes the user's behavior pattern that may be caused by some change of environments or some initiative EIR's action, understands the user's emotional condition and then acts by ingratiating the user. Finally the robot agent gets feedbacks from the user's response for proper behavior, and utilizes them as learning input of reinforcement form for the user's sake.

5.5.1 Sensing and Transformation

To observe the user's behavior and environment information, Interactive Robot has a micro head type CCD camera, ultrasonic sensor, phototransistors, temperature sensor, and humidity sensor. Transformed features from measured data are described in Table 1.

Table 1 Sensing and transformation in Interactive Robot

Purpose	Sensor	Transformed feature
Observing the user's behavior	CCD camera	Face direction
	Ultrasonic sensor	Distance to object
	Force sensor	"Hit the IR" or "Stroke the IR"
Observing the environment	Temperature sensor	Discomfort
	Humidity sensor	
	Phototransistor	Brightness

5.5.2 Classification

EIR has the neural network-based reinforcement learning architecture, shown in Fig. 12, which learns the user's personality by means of the user's satisfaction and estimates the user's emotional state with respect to user's behavior, and finally determines an action ingratiating the user.

5.5.3 Service Action

For the ingratiating action of the robot, it is designed to play three types of music: dance, ballad and rock music, or nothing with dance depending on user's emotional state since music is very effective to ingratiate itself with the person, make feel good, or make be comfortable.

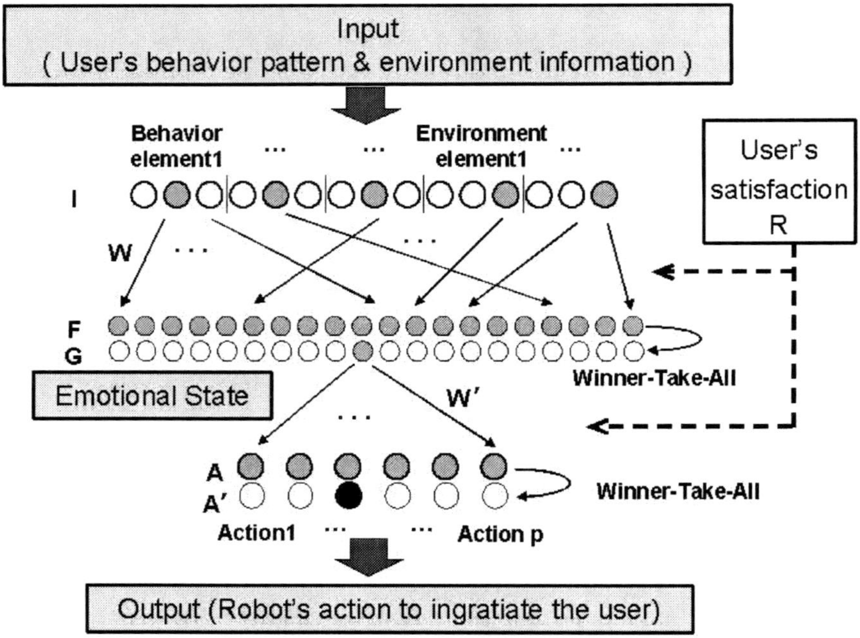

Fig. 12 Neural network-based reinforcement learning architecture of EIR

5.6 Person Identification Based on Footprint [29][30][31

As the technologies for service automation systems have been developing, new trend that requires some personalized service using learning technique is emerging. To realize this personalized service, the first step is person identification. By this reason, many diverse methods have been developing to make more natural and more unconstrained identification.

Among these, one of the most natural and unconstrained identification methods is footprint-based person identification method.

5.6.1 Sensing

Shoe-type pressure sensors are used for observing user's walking behavior. The size of this sensor is 30x10 cm^2 including 60x21 sensors on each foot. Maximum sampling time is 100 Hz. After user's one-step walking, time-series of footprint data for left and right foot are used to obtain a feature.

5.6.2 Transformation

After getting one-step walking footprint data from pressure sensors, COPs (Center Of Pressure) are extracted from each sequential walking footprint and these COPs are quantized by 4x4 to reduce the dimensionality. Finally, quantized COP trajectory is created as a feature for person recognition.

5.6.3 Classification

Using these quantized COP trajectory as an input, we configure *N* right-foot HMMs (Hidden Markov Model) and *N* left-foot HMMs for *N* user. Then, current user ID is found among *N* users by comparing the output probabilities of *2N* HMMs. In addition, NN (Neural Network) using Levenberg-Marquart learning method is used to enhance the person recognition rate from the outputs of *N* left-foot HMMs and *N* right-foot HMMs. We tested using 5 subjects with bare foot. From 14 training steps and 7 test steps for each subject, person recognition rate is 85.7%.

5.6.4 Service Action

After finding user ID, this result can be used for many personalized services of Intelligent Sweet Home such as automatic temperature control by each user's favorite temperature degree. Even when many users are in the Intelligent Sweet Home at the same time, automatic temperature control is also possible from averaging each user's favorite temperature degrees with user priorities.

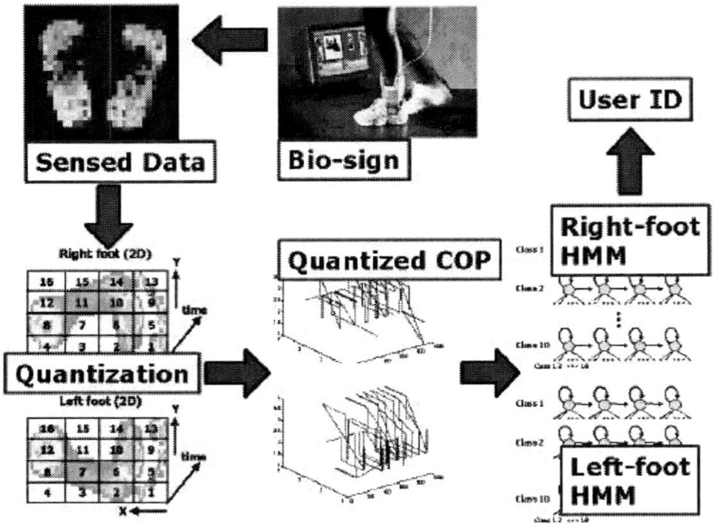

Fig. 13 Dynamic footprint-based person recognition

6 Concluding Remarks

In this paper, we have proposed an effective intention reading scheme as a means of human-robot interaction for human centered service robotic systems. By presenting several successful examples applied to service robotic systems, we have shown that the proposed scheme can deal with unconscious behavior and implicitly represented intention, and that various soft computing techniques are effective for intention reading to deal with vague and uncertain situations which can not be easily solved by conventional mathematical approach. From the results, we expect that human-robot interaction technologies with the proposed scheme would offer excellent prospects in near future.

Acknowledgement

This research is partially supported by Human-friendly Welfare Robot System Engineering Research Center, KAIST, Daejeon, Repuclic of Korea and is partially supported by the Ministry of Science and Technology of Korea as a part of Critical Technology 21 Program.

References

[1] Z. Bien and W. Zhu, "Service Robotics with Special Attention to Surgical Robots and

Rehabilitation Robotics," KITE Journal of Electronics Engineering, vol. 7, pp. 13-14, 1996

[2] W. Yu and Z. Bien, "Design of Fuzzy Logic Controller with Inconsistent Rule Base," Journal of Intelligent Fuzzy Systems, vol. 2, pp. 147-159, 1994

[3] T. Sato, Y. Nishida, J. Ichikawa, Y. Hatamura, and H. Mizoguchi, "Active Understanding of Human Intention by a Robot through Monitoring of Human Behavior," Proceedings of IROS '94, pp. 405-414, 1994

[4] T. Yamaguchi, S. Mizuno, T. Yoshida, and T. Hashimoto, "Cooperative Works for Agent Robot and Human Using Robot Vision Based on the Model of Knowledge, Emotion and Intention," Proceedings of IEEE SMC '99, pp. 987-992, 1999

[5] L. J. Haasbroek, "Advanced Human-Computer Interfaces and Intent Support: a Survey and Perspective," Proceedings of SMC '93: 'Systems Engineering in the Service of Humans', pp. 350-355, 1993

[6] Y. Inagaki, H. Sugie, H. Aisu, S. Ono, and T. Unemi, "Behavior-Based Intention Inference for Intelligent Robots Cooperating with Human," Proceedings of FUZZ-IEEE '95, vol. 3, pp. 1695-1700, 1995

[7] L. A. Zadeh, "Soft Computing and Fuzzy Logic," IEEE Software, vol. 11, pp. 48-56, 1994

[8] W. –C. Bang and Z. Bien, "New Incremental Inductive Learning Algorithm in the Framework of Rough Set Theory," Journal of Fuzzy Systems, vol. 1, pp. 25-36, 1999

[9] J. S. Jang, C. –T. Sun and E. Mizutani, Neuro-Fuzzy and Soft Computing, Prentice-Hall, Englewood, Cliffs, NJ, 1997

[10] D. –J. Kim, W. –K. Song, J. –S. Han, Z. Zenn Bien, "Soft Computing Based Intention Reading Techniques as a Means of Human-Robot Interaction for Human Centered System," Journal of Soft Computing, vol. 7, pp. 160-166, Springer-Verlag, 2003

[11] W. –K. Song, Z. Zenn Bien, "Blend of Soft Computing Techniques for Effective Human-Machine Interaction in Service Robotic Systems," Fuzzy Sets and Systems, vol.134 no.1, pp 5-25, January 2003

[12] Z. Zenn Bien, D. –J. Kim, D. H. Stefanov, J. –S. Han, H. –S. Park, P. –H. Chang, "Development of a Novel Type Rehabilitation Robotic System KARES II," Proceedings of the 1st Cambridge Workshop on Universal Access and Assistive Technology(CWUAAT 2002) , pp.201-212, Trinity Hall, Univ. of Cambridge, Cambridge, United Kingdom, March 25-27, 2002

[13] Z. Bien, P. H. Chang, W. –K. Song, and H. –S. Park, "Wheelchair-Based Robotic Arm System, KARES II, for the Disabled - a Robotic Arm and Visual Servoing," Workshop on Robotic and Automation Tools for Specific Human Assistance Tasks, IEEE ICRA, Organized by A. Bejczy and P. Fiorini, Seoul, Korea, May 22, 2001

[14] Z. Bien, W. –K. Song, D. –S. Kwon, M. –J. Chung, P. H. Chang, H. –S. Park, D. –J. Kim, J. –H. Kim, K. Lee, "A Wheelchair Robot System and Its Various Interface Methods for the Disabled Persons," The First Workshop on Technical Challenge for Dependable Robots in

Human Environments, (International Advanced Robotics Programme, IEEE Robotics and Automation Society), Seoul, Korea, May 21-22, 2001

[15] Z. Zenn Bien, K. –H. Park, W. –C. Bang and D. H. Stefanov, "LARES: an Intelligent Sweet Home for Assisting the Elderly and the Handicapped," Proceedings of the 1st Cambridge Workshop on Universal Access and Assistive Technology, Cambridge, UK, pp. 43-46, 2002

[16] G. –T. Park and Z. Bien, "Neural Network-Based Fuzzy Observer with Application to Facial Analysis," Pattern Recognition Letters, vol. 21, pp. 93-105, February 2000

[17] W. –K. Song, D. –J. Kim, J. –S. Kim, and Z Bien, "Visual Servoing for a User's Mouth with Effective Intention Reading in a Wheelchair-Based Robotic Arm," Proceedings of the 2001 IEEE International Conference on Robotics & Automation, Seoul, Korea, May 21-26, pp. 3662-3667, 2001

[18] D. –J. Kim and Z. Bien, "Soft Computing-Based Intention Reading through the User's Mouth for Human-friendly Human-Robot Interaction," Proceedings of SCIS & ISIS 2002, 23-Q3-5, 2002

[19] J. V. Basmajian and C. J. De Luca, Muscle Alive, Williams & Wilkins, 5th Edition, 1985

[20] J. –S. Han, W. –C. Bang, and Z. Zenn Bien, "Feature Set Extraction Algorithm Based on Soft Computing Techniques and Its Application to EMG Pattern Classification," Fuzzy Optimization and Decision Making, vol. 1, no. 3, pp. 269-286, August, 2002

[21] J. –S. Han, D. H. Stefanov, K. –H. Park, H. –B. Lee, D. –J. Kim, W. –K. Song, J. –S. Kim, and Z. Zenn Bien, "Development of an EMG-Based Powered Wheelchair Controller for Users with High-Level Spinal Cord Injury," International Conference on Control, Automation and Systems, Jeju, Korea, pp. 503-506, October 17-21, 2001

[22] Z. Zenn Bien, J. –B. Kim, D. –J. Kim, W. Jang, "Issues of Sign Language Recognition for Visual Communication," Japan Korea Joint Workshop on Intelligent Human Friendly Mechatronics and Systems, Tokyo, Japan, pp. 60-63, November 3-5, 1999

[23] J. –B. Kim, Z Bien, W. –C. Bang, "Intelligent Recognizer of Compound Words in the Continuous Korean Sign Language," Fourth World Automation Congress(WAC2000), Hawaii, USA, June 11-15, 2000

[24] J. -B. Kim, K. –H. Park, W. –C. Bang and Z. Zenn Bien, "Continuous Gesture Recognition System for Korean Sign Language Based on Fuzzy Logic and Hidden Markov Model," Proceedings of the 2002 IEEE International Conference on Fuzzy Systems(FUZZ-IEEE2002), vol. 2, pp. 1574-1579, 2002

[25] H. Jang, D. –J. Kim, J. –B. Kim and Z. Bien, "A Study on Hand-Signal Recognition System in 3-Dimensional Space," Proceedings of IEEK Summer Conference, vol. 3, pp.215-218, 2002

[26] Z. Zenn Bien and J. –H. Do, "Interactive Robot for Emotion Monitoring," Korea-Japan Joint Workshop on Network Based Human Friendly Mechatronics and Systems, Seoul, Korea,

pp. 62-65, November 16-18, 2000

[27] J. –H. Do, K. –H. Park, Z. Bien, K. C. Park, Y. T. Oh and D. H. Kang, "A Development of Emotional Interactive Robot," The 32nd International Symposium on Robotics, Seoul, Korea, pp. 544-549, April, 2001

[28] Z. Zenn Bien and J. –H. Do, "An Interactive Robot with Emotion Monitoring Capability," 1st Korea-China Joint Workshop on Robotics, Pohang, Korea, pp. 44-49, May 27-28, 2001

[29] J. –W. Jung, T. Sato, and Z Bien, "Unconstrained Person Recognition Method Using Dynamic Footprint," Proceedings of SCIS & ISIS 2002, 22-Q4-2, 2002

[30] J. –W. Jung, K. –H. Park, and Z. Bien, "Unconstrained Person Recognition Method Using Static and Dynamic Footprint," Proceedings of the Eighteenth Hungarian-Korean Seminar, pp. 129-137, 2002

[31] J. –W. Jung, T. Sato, and Z. Bien, "Unconstrained Person Recognition Method Using Dynamic Footprint," Proceedings of International Conference on Fuzzy Information Processing 2003, vol. II, pp. 531-536, 2003

Farbenbasierte Objektdetektion in der Service-Robotik

Ivan Volosyak
Institut für Automatisierungstechnik
Universität Bremen
Otto-Hahn-Allee NW1, D-28359 Bremen
volosyak@iat.uni-bremen.de

Kurzfassung

In diesem Beitrag wird das Problem der Identifikation von Gegenständen anhand der Objektfarbe vor allem für Aufgaben der Service-Robotik behandelt. Der Einsatz des Erkennungssystems in dem visuellen Regelkreis eines Roboterarms erfordert extrem schnelle Verarbeitungszeiten und hohe Genauigkeit bei der Objektlokalisierung im Bild. Um diesen Anforderungen zu entsprechen, werden die natürlichen Eigenschaften der zu detektierenden Objekte verwendet. Unter einer natürlichen Markierung wird hierbei der Gegenstand verstanden, bei dem seine Farbe ihn genau charakterisiert und eine Zuteilung zu einer Sorte von Bildobjekten ermöglicht. Zur Objekterkennung werden die gewöhnlich zeitaufwendigeren Methoden aus der Farbbildverarbeitung verwendet; die Beschleunigung der Verarbeitung wird mittels eines neuen fuzzybasierten Farbklassifizierungsverfahrens realisiert. Die Robustheit der Objektdetektion wird dadurch gewährleistet, dass Farbverteilungen der HS-Kanäle des HSV (Hue-Sättigung-Value)- Farbraums für einfarbige Gegenstände praktisch beleuchtungsinvariant sind und eine fuzzybasierte Klassifizierung stattfindet. Zur Parametereinstellung wurde ein separates Programm implementiert, das eine schnelle Systemanpassung an neue Objekte ermöglicht.
Die erzielten Ergebnisse wurden mit dem Rehabilitationssystems „FRIEND“ getestet.

1 Einleitung

Der am IAT entwickelte Rehabilitationsroboter „FRIEND“ dient der Unterstützung von behinderten Personen. Typische Abläufe wie das Greifen eines ausgewählten Gegenstandes oder das Einschenken eines Glases werden mit einem visuell geregelten Roboterarm realisiert. Für diese Aufgabe müssen verschiedene Objekte sicher erkannt werden [1]. Die visuelle Regelung des Roboterarms erfordert die Verarbeitungszeiten in einem zulässigen Bereich zu halten und aus Sicherheitsgründen eine Framerate von mehr als 10 fps (frames per second) zu realisieren. Um diese Verarbeitungsgeschwindigkeit zu erreichen, werden die Objekte üblicherweise künstlich markiert [2]. Mit zunehmender Leistungsfähigkeit der Bildverarbeitungs- Hard- und Software ist es mittlerweile möglich geworden, zur Problemlösung von der „einfacheren“ Grauwert- und Schwarz/Weiss- Bildverarbeitung zur Farbbildverarbeitung sowie Stereobildverarbeitung überzugehen. Grundsätzlich sind durch die Nutzung von Farbbildern zur Objektdetektion qualitative Verbesserungen zu erwarten. Eine große Schwierigkeit ist dabei, dass die Objektfarbe nicht exakt, sondern nur vage definiert werden kann. Der Grund dafür sind große Schwankungen in der Farbe eines Objektes, die durch störende Einflüsse wie Umgebungsbeleuchtung, Schattenwurf, Lichtreflexe u.s.w. verändert werden kann. Auch bei der Aufnahme einer Bildsequenz können Farbabweichungen von Bild zu Bild auftreten. Eine robuste Objektlokalisierung im Fall mehrerer zu detektierender Gegenstände, die in praktisch jedem Rehabilitationsszenario notwendig ist, erweist sich dabei als sehr schwierig, ist aber eine der unverzichtbaren Voraussetzungen für die Funktionsweise des Robotersystems. Eine Vereinfachung wird dadurch erreicht, dass nur die für die typischen Abläufe relevanten Objekte zu lokalisieren sind.

2 Möglichkeiten zur farbenbasierten Klassifikation

Aufgrund der angeforderten sehr kurzen Verarbeitungszeiten in dem visuellen Erkennungssystem scheiden praktisch sämtliche gängigen Farbbildsegmentierungsverfahren, die auf einer vollständigen Bildsegmentierung basieren, sofort aus. Der Grund dafür liegt darin, dass die durchschnittliche Zeit dieses Vorganges deutlich über den maximal erlaubten 100 ms liegt. Es wurde z.B. das Segmentierungsverfahren, basierend auf CSC (Color Structure Code [3]) reimplementiert; bei einer Bildgröße von 512 x 256 Pixel dauerte alleine die Segmentierung auf einem 2,8 GHz Pentium IV über 140 ms. In [4] wurde bei dem Einsatz dieses Verfahrens auf die Verarbeitungszeiten unter 2 fps hingewiesen. Es wurden auch eine Reihe anderer Farbsegmentierungsverfahren analysiert; einen guten Überblick findet man z.B. in [5]. Darüber hinaus kommt zur Lösung dieser Aufgabe nur eine sehr schnelle Segmentierung, basierend z.B. auf Farbfilterung, in Frage. Parallel müssen aber viele andere Sicherheitsvorkehrungen zur Gewährleistung einer zuverlässigen Objekterkennung getroffen werden, um diese schnelle Segmentierung zu rechtfertigen.

2.1 Hardware

Die im Rahmen des Projekts „FRIEND" zur Verfügung stehende Hardware wirkt sich an diversen Stellen auf die Art der Implementierung aus. Deshalb werden die wesentlichen Hardwarekomponenten hier kurz beschrieben. Bild 1 zeigt den Aufbau des Systems. Das System besteht aus

- Rollstuhl "MEYRA"
- MANUS Roboterarm
- PC (inkl. Framegrabber, CAN-Bus.)
    - Sprachverarbeitung
    - Bildverarbeitung
    - Pan-Tilt- Controller
    - Zoom- Controller
    - MANUS- Arm- Controller
    - Controller für das intelligente Tablett
    - CAN-Bus Interface
    - Kraftmomentensensor am Greifer
- Greifermontierte Miniaturkamera
- Stereokamerasystem mit Pan-Tilts
- LCD Display am Tablett

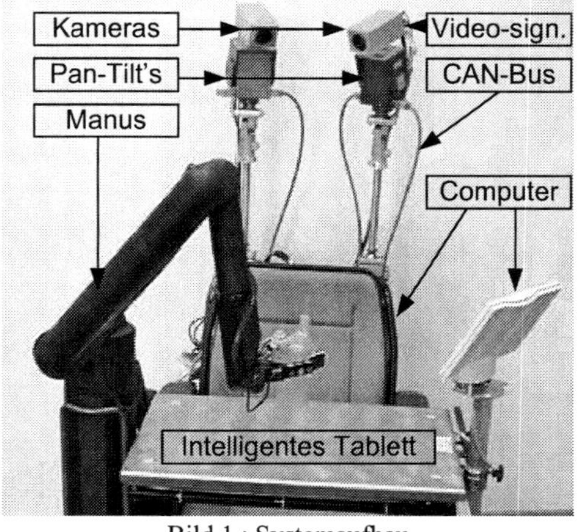

Bild 1 : Systemaufbau

Hierbei ist wichtig zu erwähnen, dass sämtliche Steuerungs- und Bildverarbeitungsaufgaben auf einem handelsüblichen Intel Pentium- basierten PC laufen.

3 Klassifizierung

Der Schwerpunkt der vorgeschlagenen Methode der automatischen Objektdetektion liegt in der neuen fuzzybasierten Klassifizierungsmethode. Der nächste Abschnitt gibt einen kurzen Überblick über das Klassifikationsproblem allgemein und die gängigen Lösungsansätze.

3.1 HSV Farbraum

Die Intensität der Farben variiert bei der Bildaufnahme im Fall einer nicht konstanten Umgebungsbeleuchtung und beim Schattenwurf. Diese Abhängigkeit ist oft unerwünscht. Um sie auszublenden, wird in der Literatur meist eine Auftrennung eines Farbbildes in einen chromatischen und einen achromatischen Anteil vorgeschlagen. Diese Auftrennung ist in einem HSV- Farbraum bereits vorhanden, wodurch sich die in diesem Farbraum arbeitenden Segmentierungsverfahren als robust erweisen [6]. Da der Rechenaufwand dieser Farbraumtransformation sehr gering und in vielen neuen Framegrabbern bereits auf der Hardware implementiert ist, wurde der HSV- Farbraum zur Segmentierung ausgewählt.

3.2 Threshold Klassifizierung

Bei der Klassifizierung wird jedes Pixel anhand seiner Farbwerte zu einer Farbklasse zugeordnet. Um den in der Einleitung beschriebenen Anforderungen an die Verarbeitungszeiten unter 100 ms zu entsprechen, können in dem System nur schnelle Klassifizierungsverfahren eingesetzt werden. Die einfachste Klassifizierungsmethode benutzt für die Farbklassenzuordnung eine Reihe von vorher bestimmten Threshold- Werten. Die Farbklassen werden in diesem Fall durch eine entsprechende Partitionierung des Farbraums definiert, was im Fall einer äquidistanten Einteilung der einzelnen Farbkanäle in Bild 2 veranschaulicht ist.

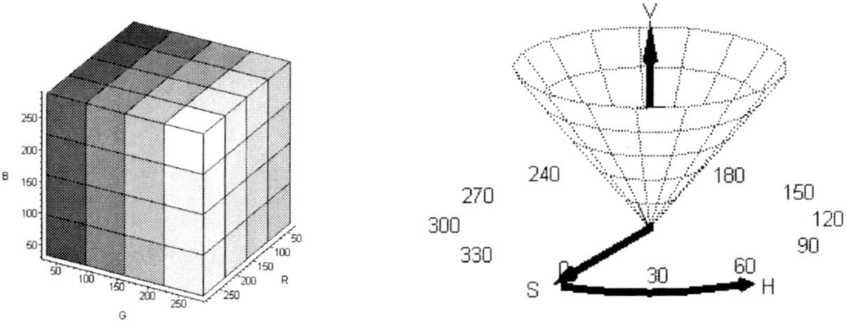

Bild 2 : Definition der Farbklassen in RGB und HSV Farbräumen

Da die Objektfarben nicht das gesamte Farbspektrum abdecken, werden nicht alle Farbwerte zu Objektklassen zugeordnet. Das Problem ist jedoch, dass im Fall mehrerer zu detektierender Gegenstände ihre Farbklassen praktisch immer eine Tendenz zur Überschneidung aufweisen. Ein großer Nachteil ist dabei, dass die Pixel mit Farbwerten, die nahe an einer Klassengrenze

liegen, durch eine Beleuchtungsänderung oder selbst durch geringes Rauschen zu einer anderen Farbklasse zugeordnet werden können. Daher sind die Objekterkennungssysteme, in denen lediglich eine feste Klassendefinition verwendet wird, sehr unzuverlässig.
Um diese Probleme zu überwinden, wurde eine fuzzybasierte Klassifizierungsmethode entwickelt, die nachstehend detailliert beschrieben ist.

4 Fuzzy - Einsatz

Bei der Benutzung von konstanten Threshold- Werten werden diese größtenteils entweder manuell ausgewählt oder nach einer Berechnung von Mittelwert und Varianz [7] für die entsprechenden Histogramme der HSV- Kanäle. Für einfachere Handhabung wurde für das System FRIEND ein Belehrungstool (ein separates Programm) implementiert, das eine sehr schnelle Anpassung der Farbklassen ermöglicht. Das Tool bietet dem Benutzer ein Graphical-User-Interface (GUI), womit eine manuelle Auswahl des Bereichs, der ausschließlich den gewünschten Gegenstand beinhaltet, möglich ist. Für diese Region werden Histogramme der HSV- Farbkanäle berechnet. Dadurch wird eine manuelle Definition der Threshold- Werte ermöglicht, s. Bild 3.

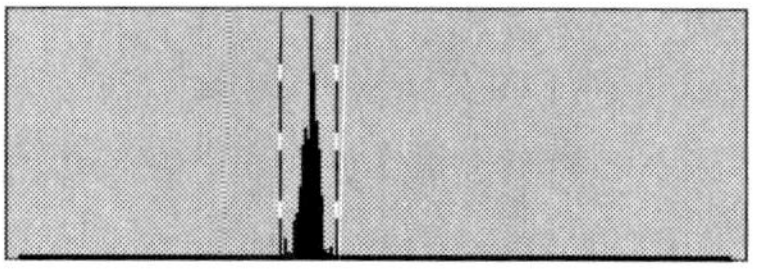

a) Histogramm im Hue-Kanal

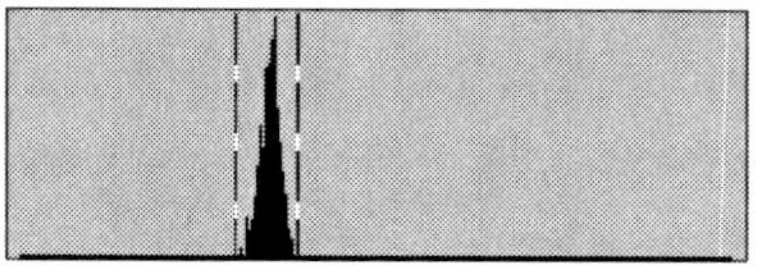

b) Histogramm im Sättigungskanal

Bild 3 : Histogramme der Objektfarben

Die aus diesem Vorgang ermittelten Grenzwerte definieren den Farbbereich, in dem sich befindende Pixel mit einer 100% Wahrscheinlichkeit zu der jeweiligen Klassenregion gehören. Alle Pixel mit Werten innerhalb der definierten Grenzen werden der jeweiligen Klasse zugeordnet. In diesem 1. Schritt werden die Histogramme bei einer spezifischen Beleuchtung und Lage bestimmt. Um zu enge Grenzwerte zu vermeiden, werden die Beleuchtung und Lage künstlich variiert und die Überlagerung der Histogramme zur Definition der Grenzwerte herangezogen. Diese Maßnahmen verbreitern den Farbbereich, der eine Klasse definiert. Die Änderung der Objektlage ist notwendig, da sie im Vergleich zur Beleuchtung wegen starken Reflexionen sogar eine größere Auswirkung auf die Objektfarben hat. Alle Farbwerte, die außerhalb dieses erweiterten Farbbereichs liegen, gehören nicht zu dieser Objektklasse. Für den Umgang mit den bei diesem Vorgang gewonnen 4 Werten wurde zur Farbklassendefinition ein Fuzzy- Ansatz mit linguistischen Variablen, die den Farbkanälen entsprechen, gewählt. Die Fuzzy- Zugehörigkeitsfunktion kann in der Trapezform abgebildet werden, wie es in Bild 4 dargestellt ist. Dass die Flanken gleiche Steilheit aufweisen, ist hier nicht notwendig und auch eher unwahrscheinlich. Diese Realisierung erlaubt eine unscharfe Aussage über die Pixelzugehörigkeit zu treffen: Z.B. das Pixel gehört mit 55% zu Hintergrund und mit 45% zu der Farbklasse 2. Dies macht das Verfahren robuster, da über die Auswertung z.B. der Nachbarnpixel eine bessere Entscheidung über die Pixelzugehörigkeit getroffen werden kann.

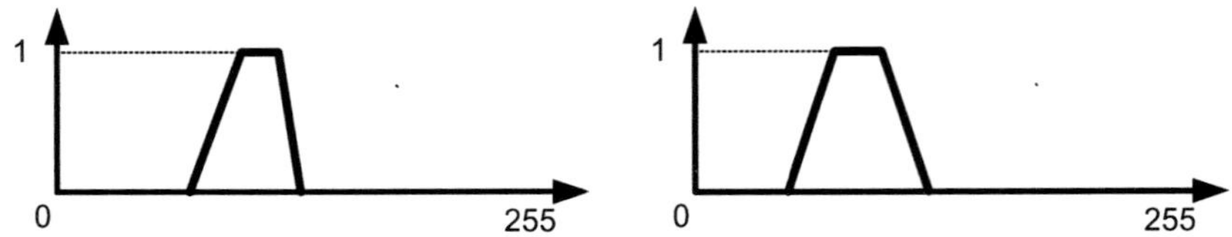

a) Unscharfe Definition einer Farbklasse, HUE- Kanal

b) Unscharfe Definition einer Farbklasse, Sättigungskanal

Bild 4 : Fuzzy Zugehörigkeitsfunktionen

Diese Prozedur wird für alle Objekte wiederholt. Die daraus gewonnene Fuzzy-Zugehörigkeitsmenge kann über die mit dem Hintergrund korrespondierende Zugehörigkeitsfunktion erweitert werden. Die Entscheidung über die Pixelzugehörigkeit wird mittels Produktberechnung der den Farbkanälen entsprechenden Zugehörigkeitsfunktionen getroffen. Bei Überschneidungen der Fuzzy- Zugehörigkeitsfunktionen müssen diese evtl. korrigiert werden, falls keine eindeutige Klassifizierung durchzuführen ist. Auf die konkrete Art der Korrektur wird hier nicht näher eingegangen, eine Möglichkeit ist [12] zu entnehmen.
Die beschriebene Fuzzy- Klassifizierung kann in zwei Ausführungen verwendet werden:

- Einstufige Methode, bei der ein Pixel sofort der Farbklasse zugeteilt wird, zu der es mit größter Wahrscheinlichkeit gehört (die Fuzzy- Menge soll diese eindeutige Zuteilung ermöglichen, ggf. nach der beschriebenen Korrektur);
- Zweistufige Methode, zuerst wird das klassische Verfahren angewendet (entweder 100% Bereich oder die volle einstufige Methode), danach wird z.B. nur für geometrisch abgegrenzte Bereiche eine zweite Klassifizierung mit verbreiteten Grenzwerten durchgeführt. Diese Methode kann mittels Rückkopplungsstrukturen erweitert werden.

5 Praktische Realisierung

Das entwickelte Verfahren wurde für den Rehabilitationsroboter FRIEND implementiert. Dieser Abschnitt gibt eine Auskunft über die vorhandene Systemrealisierung.

5.1 Systemaufbau

Den prinzipiellen Aufbau des Erkennungssystems zeigt Bild 5. Die obere Reihe entspricht der Standardaufteilung der Verarbeitungsschritte in einem technischen Bildverarbeitungssystem. Darunter wird die verwendete Realisierung anhand eines Beispiels „Erkennung einer Flasche" dargestellt.

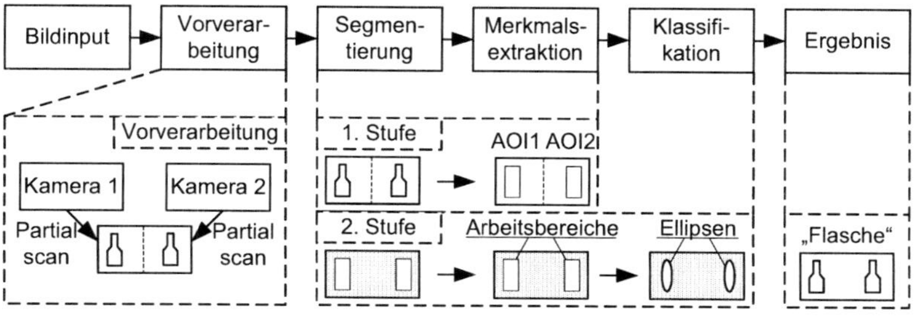

Bild 5 : Prinzipieller Aufbau des Erkennungssystems

5.2 Stereokamerasystem

Die visuelle Regelung des Roboterarms erfordert die Bildinformationen eines Stereokamerasystems. Dieses besteht im FRIEND- System aus zwei SONY EVI 371 PAL Farbkameras. Die parallele Aufnahme der „online“ - Farbbilder im PAL-Format von zwei Kameras in einem Rechner verursacht grundsätzlich eine fehlerhafte Bildübertragung durch eine sehr hohe PCI- Bus- Belastung (einige Bildzeilen fallen aus, Ursachen: Hoher Datentransfer, 2 Framegrabber, die angeforderte Framerate). Als eine mögliche Lösung zur Reduktion des Datentransfers wurde die Arbeit in einem noninterlaced Modus ausgewählt; dabei werden nur die Halbbilder übertragen (mit einer Größe 768 x 288 Pixel anstatt der Standard PAL- Auflösung von 768 x 576 Pixel).
Bei der Verwendung der Multikamerasysteme tritt hauptsächlich das Problem der Synchronisation der Verarbeitungsschritte auf. Dieses Problem lässt sich durch eine rationelle Verbindung der Eingangsbilder sofort nach der Bildaufnahme lösen. Nach der Bildaufnahme in beiden Kameras findet zuerst eine Partial- Scan- Methode je Bild statt, dabei wird aus jedem aufgenommenen Bild der benötigte Bereich mit der Größe von 512 x 256 Pixel ausgeschnitten; als Ergebnis wird das Bild der Größe 256 x 256 Pixel mit normalen Seitenverhältnissen ausgegeben. Mit dem Bild der zweiten Kamera geschieht dasgleiche. Im zweiten Schritt werden beide Bilder zu einem Bild der Größe 512 x 256 verknüpft. Bild 6 zeigt den angewendeten Verknüpfungsprozess.

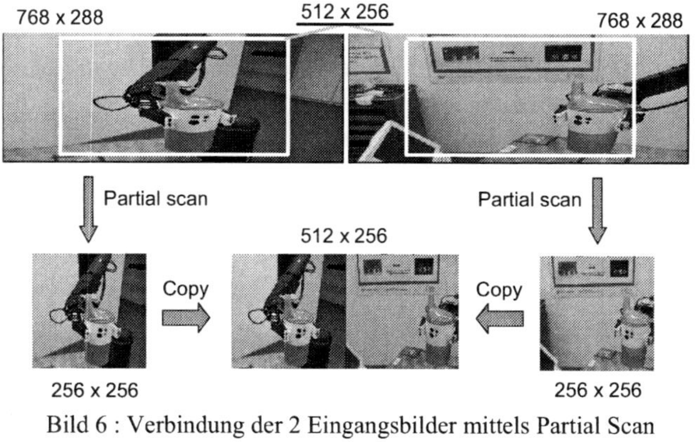

Bild 6 : Verbindung der 2 Eingangsbilder mittels Partial Scan

Bisher wurde diese Methode noch nicht in der Farbbildverarbeitung aufgrund der Datentransferprobleme verwendet, dieser Ansatz zeigt eine gute Verwendbarkeit anhand der praktischen Realisierung.

5.3 Vorverarbeitung

Die Vorteile des HSV- Farbraums wurden bereits beschrieben. An dieser Stelle findet eine Konvertierung des Bildes aus dem von der Kamera verwendeten RGB- Farbraum zum HSV-Farbraum statt. Bild 7 zeigt außerdem den Effekt des Farbrauschens.

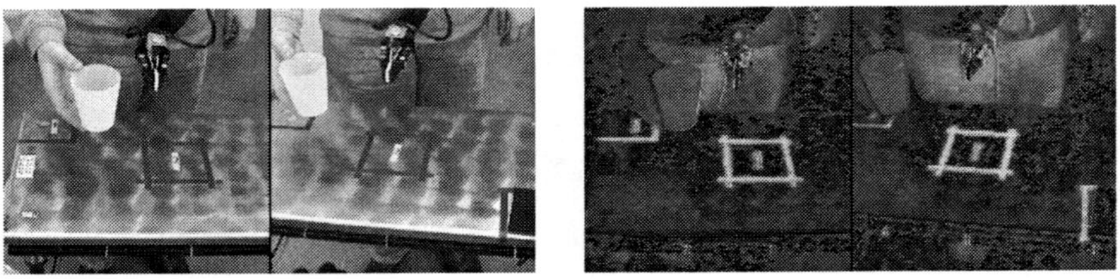

Bild 7 : RGB → HSV- Farbraumkonvertierung

5.4 Erste (fuzzybasierte) Klassifizierung

Nach der Vorverarbeitung findet eine erste Segmentierung statt, in der alle Pixel anhand ihrer Farben zu Klassenregionen zugeordnet werden.
Dafür wird eine der zwei vorgestellten Methoden benutzt: Entweder die klassische Methode mit konstanten Thresholds, bei der nur die Pixel ausgewertet werden, die mit 100% Wahrscheinlichkeit zu der jeweiligen Klassenregion gehören, oder es wird das vorgeschlagene fuzzybasierte Verfahren benutzt. Dabei wird die Wahrscheinlichkeit, dass das Pixel zu der Farbklasse gehört, durch das Produkt seiner Zugehörigkeitsfunktionen berechnet. Es entsteht ein binäres Bild pro Klasse, das sogenannte Kristallisationspunkte für die jeweiligen Farbregionen beinhaltet. Jeder Farbklasse wird bei dieser Vorgehensweise ein binäres Bild zugewiesen, s Bild 8. Die Objektidentifikation findet bereits an dieser Stelle statt, die nächsten Verarbeitungsschritte dienen hauptsächlich der Positionsbestimmung, letztens findet eine Plausibilitätskontrolle statt. Bild 9 zeigt außerdem, dass die Ergebnisse der ersten Klassifizierung auf keinen Fall völlig fehlerfrei sind.

Die Hauptaufgabe der nachgeschalteten Verarbeitungsschritte besteht in der Unterdrückung der Ausreißer (s. 5.5) und in der Zuteilung von der maximal möglichen Anzahl der Objektpixel zu den Objektregionen (s. 5.6). Je mehr Pixel richtig segmentiert werden, desto präziser werden die Ergebnisse der Objektlokalisation (s. 5.7). Es hat sich herausgestellt, dass eine alleinige Definition der Objektlage mittels umschreibenden Rechtecke (s. 5.5) unausreichende Daten für die nachfolgenden Visual- Servoing- Aufgaben zurückliefert, da vor allem die Informationen über die relative Lage (Winkel) fehlen.

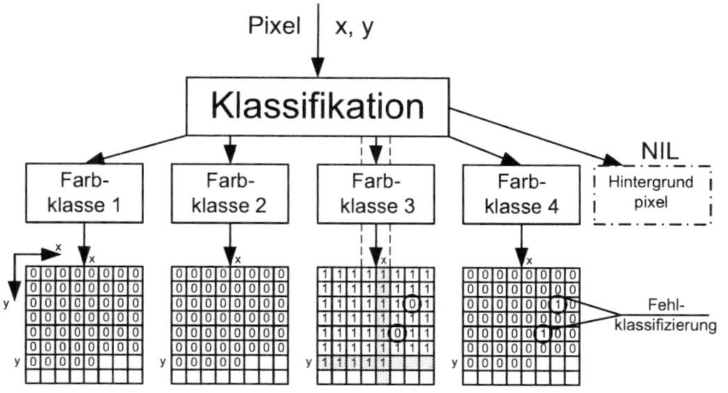

Bild 8 : Angewendete Klassifizierung

Bild 9 macht diesen Verarbeitungsschritt anhand eines realen Bildes deutlich.

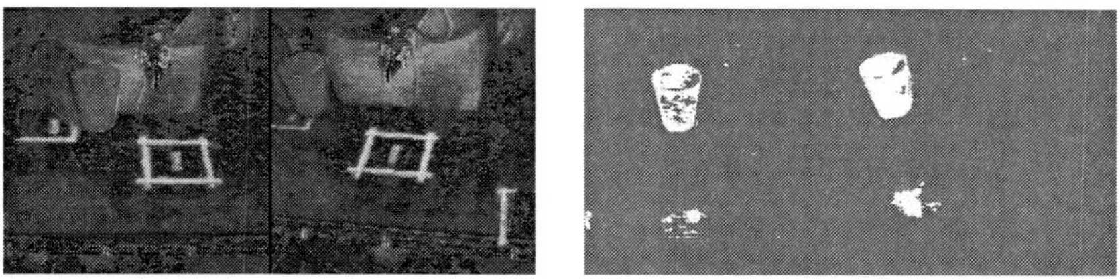

Bild 9 : Binärbild als Ergebnis der ersten Klassifizierung

In diesem binären Bild sind auch Pixel, die durch Rauschen, Reflexionen u.s.w. ähnliche Farbwerte besitzen, (als Ausreißer) deutlich zu sehen.

5.5 Erste Merkmalsextraktion

Die erste Merkmalsextraktion erfolgt mit der Bildverarbeitungssoftware CVB Blob [10]. Diese fasst die aneinandergrenzenden Pixel zu Objekten, den sogenannten Blobs, zusammen. Als Ergebnis werden zwei umgrenzende Rechtecke pro Farbklasse zurückgegeben. Dieser Prozess ist dem Bild 10 zu entnehmen.

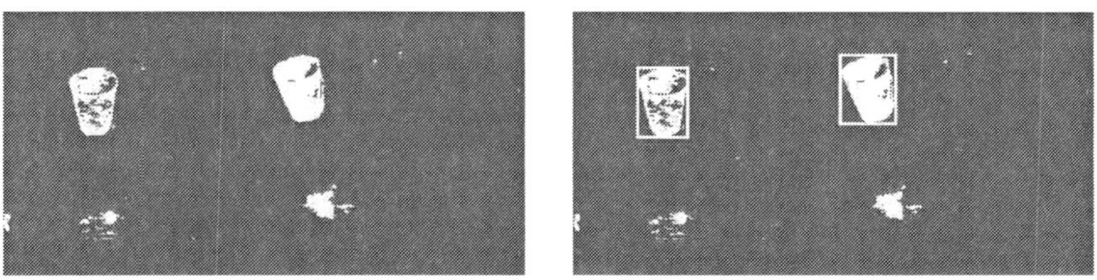

Bild 10 : Erste Merkmalsextraktion als Ausgang 2 umschreibenden Rechtecke

5.6 Zweite fuzzybasierte Klassifikation

Bei der zweiten Farbklassifikation werden nur die Pixel des Originalbildes ausgewertet, die geometrisch innerhalb des umschreibenden Rechtecks (oder in einem gegebenen Toleranzmaß, z.B. Höhe ± 20%, Breite ± 10% [9]) liegen und zum vollen Farbbereich der jeweiligen Farbklasse gehören (für diese Pixel ist der Produkt der Zugehörigkeitsfunktionen größer Null; unabhängig von der in der ersten Klassifizierung gewählten Methode werden hierbei die Fuzzy Zugehörigkeitsfunktionen verwendet). Je mehr zu der Region gehörenden Pixel richtig extrahiert werden, desto höher ist die Qualität der nachgeschalteten Verarbeitungsstufe. Das Verfahren bleibt stabil, da die Objektbereiche über Rechtecke bereits fest definiert sind. Diesen Verarbeitungsschritt zeigt Bild 11.

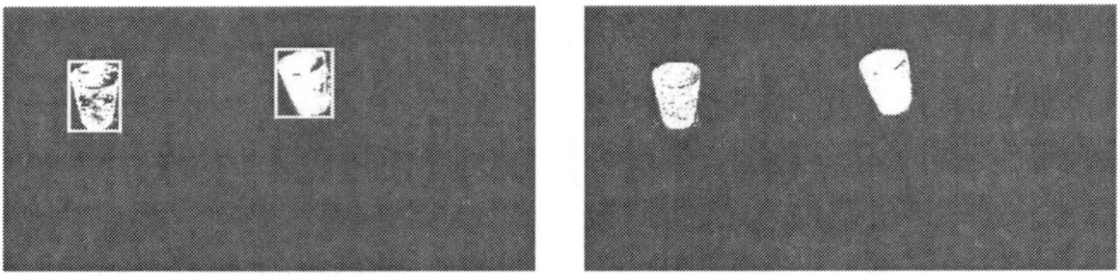

Bild 11 : Zweite Klassifikation als Ausgang 2 Objektregionen zur Ellipsenanpassung

5.7 Endgültige Merkmalsextraktion

Die zu lokalisierenden Objekte befinden sich in den zuvor bestimmen umschreibenden Rechtecken. Alle Pixel, die sich geometrisch in den Rechtecken befinden und gemäß deren Farbwerten in dem verbreiterten Farbbereich der jeweiligen Farbklasse liegen, werden zu Objektregionen zusammengefasst. Für diese Regionen werden die Momente bis zur zweiten Ordnung [11] bestimmt. Zur Objektdetektion werden Methoden der Ellipsenanpassung [8] eingesetzt, die aus diesen Momenten eine entsprechende Ellipse (mit den äquivalenten normalisierten Zentralmomenten zweiter Ordnung) bilden. Die endgültige Klassifizierung beinhaltet eine Plausibilitätskontrolle durch Vergleich entweder der Verhältnisse der Haupt- zur Nebenachse der Ellipsen oder der absoluten Größen der Ellipsen.

6 Ergebnisse

Durch Einsatz des vorgeschlagenen fuzzybasierten Farbklassifizierungsverfahrens lassen sich die beschriebenen Probleme gängiger Klassifizierungsmethoden umgehen und gleichzeitig die Zuverlässigkeit der Objekterkennung erhöhen. Bild 12 zeigt die erzielten Ergebnisse im Fall eines Szenarios aus der Rehabilitationsrobotik: „Einschenken eines Glases mit dem Roboter“. Die über die detektierten Objekte gelegten Ellipsen zeigen eine gute Anwendbarkeit des Verfahrens für Visual Servoing- Aufgaben.

Bild 12 : Ergebnisse der Objektdetektion

Die Klassifizierungsmethode wurde in Multithreading- Ausführung unter MS VC++ 6.0 SP5 implementiert. Dadurch wurde eine Beschleunigung der Verarbeitung im Fall einer kleinen Anzahl der zu detektierenden Objekte garantiert. Anhand einer Stichprobe aus 5 Gegenständen wurden folgende Ergebnisse erzielt (Tabelle 1):

Anzahl der zu detektierenden Objekte	Zeitbedarf der Farbbildverarbeitung, ms
1	79.4, entspricht ca. 12.6 fps
2	79.8, entspricht ca. 12.5 fps
3 (Einschenkszenario)	**80.1, entspricht ca. 12.5 fps**
4	112, entspricht ca. 8.9 fps
5	114, entspricht ca. 8.8 fps

Tabelle 1 : Erkennungsleistung des Systems

Der große Anstieg der Verarbeitungszeit bei der Erkennung von mehr als 3 Objekten ist Hardware- bedingt: Einige von Framegrabbern gelieferten Bilder werden nicht mehr verarbeitet, da noch eine Bearbeitung des vorhergehenden Bildes stattfindet. Diese Einbuße ist unvermeidbar, da derzeit keine Triggerung der Kameras in dem Stereokamerasystem existiert; dies stellt aber kein Problem dar, da in derzeitiger Realisierung lediglich bis 3 Gegenstände gleichzeitig zu lokalisieren sind.

7 Zusammenfassung und Ausblick

In diesem Beitrag wurde ein neues Verfahren zur automatischen Objekterkennung auf Basis der Grundfarbe der Objekte zum Einsatz in der Service- Robotik vorgeschlagen. Die wichtigsten Anforderungen an das zu entwickelnde System waren:

- Verwendung von natürlichen Eigenschaften der Objekte zur Detektion, keine künstlichen Markierungen;
- Eine Verarbeitungszeit unter 100 ms pro Verarbeitungsschritt;
- Eine robuste und genaue Objektdetektion bei unterschiedlichen Beleuchtungsbedingungen, die eine nachfolgende Datenauswertung zum Einsatz in Visual Servoing- Aufgaben des Roboterarms sicherstellt.

Das vorgeschlagene System entspricht den aufgelisteten Anforderungen. Da grundsätzlich nur die Farbinformationen über die zu detektierenden Objekte zur Erkennung benötigt werden, ist die Erkennung von teilweise verdeckten Gegenständen möglich; durch den Einsatz der Ellipsenanpassung werden dabei auch gut verwendbare Ergebnisse für die Positionsbestimmung erreicht. Bei hohem Rauschen verringert sich allerdings die Zuverlässigkeit der Methode besonders im Fall der Lokalisation von kleineren Objekten. Ein anderes Problem stellt ein farbiger Hintergrund dar, dabei zeichnen sich in den meisten Fällen die Objektfarben nur sehr undeutlich vor dem Hintergrund ab, und die der Objektkanten

benachbarten Hintergrundpixel werden als zum Objekt gehörend segmentiert. Aus diesem Verhalten folgt eine Ungenauigkeit in der Ellipsenanpassung. Zur Lösung dieses Problems ist eine nachgeschaltete Verarbeitung notwendig, die aber aus Zeitgründen (CPU Zeitbedarf) in diesem System nicht mehr bearbeitet werden konnte.
Diese Methode ist natürlich auch zur Objekterkennung allgemein gut einsetzbar. Als ein mögliches Anwendungsgebiet kann z.B. der Einsatz für fußballspielende Roboter betrachtet werden, wenn allen verwendeten Gegenständen unterschiedliche Farben zugeteilt sind. Der Fuzzy- Einsatz ermöglicht eine unscharfe Aussage über die Pixelzugehörigkeit treffen; das ermöglicht eine bessere Klassenzuordnung. Durch eine Einbeziehung der Pixelnachbarschaft und von vorhandenem Vorwissen über die zu erkennenden Objekte lässt sich eine höhere Zuverlässigkeit der Objektdetektion erreichen. Die Verwendung von Look-Up-Tabellen kann das Verfahren beschleunigen, so dass kein Unterschied im Zeitverhalten in Vergleich zu den gängigen Methoden vorliegt. Als eine mögliche Erweiterung der Methode kann die Anwendung von Rückkopplungsstrukturen in der Farbbildverarbeitung genannt werden, was z.B. in [12] vorgestellt wurde.

8 Literatur

[1] Gräser, A.; Martens, C. "**Rehabilitation Robots - Transfer of Development and Research Results to Disabled Users**" 3rd International Workshop on Human-friendly Robotic Systems, Daejeon, Korea, January 21-22, 2002.

[2] Martens, C.; Gräser, A.: "**Robust Mark-Detection for the Realization of a Robot Gripping Action**", International Conference on Multisensor Fusion and Integration for Intelligent Systems (MFI 2001), August 20th - 22nd 2001, Baden-Baden, Germany.

[3] V. Rehrmann : "**Stabile, echtzeitfähige Farbbildauswertung**", In: Koblenzer Schriften zur Informatik; Bd. 1. -Koblenz, Germany : Fölbach, 1994.

[4] I. Scholz, J. Schmidt, H. Niemann : "**Farbbildverarbeitung unter Echtzeitbedingungen in der Augmented Reality**", In: 7. Workshop Farbbildverarbeitung, Seiten 59-65. Erlangen, Germany, 4.-5. Oktober 2001.

[5] W. Hafner : "**Segmentierung von Video-Bildfolgen durch adaptive Farbklassifikation**", Dissertation, Technische Universität München, 1998.

[6] M. Ross : "**Statische Fuzzy Farbhistogramme**", In: 8. Workshop Farbbildverarbeitung, Seiten 79-86. Ilmenau, Germany, 10.-11. Oktober 2002.

[7] J. Brusey; L. Padgham: "**Techniques for Obtaining Robust, Real-Time, Colour-Based Vision for Robotics**", IJCAI-99 Proceedings of the Third International Workshop on Robocup, Stockholm, Sweden, July 1999.

[8] R. Haralick and L. Shapiro: "**Computer and Robot Vision**" Addison-Wesley Publishing Company, Reading, MA, USA, 1992.

[9] B. Frei, M. Hund, M. Schnitzlein : "**Umfeldorientierte Farbsegmentierung auf Basis eines Fuzzy- basierten Farbbelehrungstools**", In: 8. Workshop Farbbildverarbeitung, Seiten 95-109. Ilmenau, Germany, 10.-11. Oktober 2002.

[10] Stemmer Imaging GmbH: "**CVB Blob DLL – Version 3.0**", 2002

[11] W. Abmayr: "**Einführung in die digitale Bildverarbeitung**", Teubner Verlag, Stuttgart, 1994.

[12] I. Volosyak, A. Gräser. : "**Automatic object recognition using fuzzy decision system for the reha-robot FRIEND**", Proceedings of the ICORR 2003 (The Eighth International Conference on Rehabilitation Robotics), 22. - 25.04.2003, Daejeon, Korea.

Generation of Parallel Executable Control Sequences for Rehabilitation Robotic Systems on the Basis of Hierarchical Petri Nets

Christian Martens
University of Bremen, Institute of Automation
Otto-Hahn Allee, NW1, 28359 Bremen
E-Mail: cmartens@iat.uni-bremen.de

Abstract

Rehabilitation robots were designed to support disabled persons in daily life situations as well as in the working environment. Current manipulator based systems offer services on a relatively low level of abstraction, like controlling the gripper or the joints of the arm directly. This puts a high cognitive load on their users and makes their usage very inconvenient. To offer services on a higher level of abstraction, like pouring and offering a drink, becomes desirable. This requirement demands task planning capabilities. The system has to generate executable control secquences, in order to reach the goal related to the high abstraction level command. For the generation of "optimal" control sequences (e.g. in the meaning of shortest execution time), parallel executable operators have to be taken into account. This entails higher computational effort and an intelligent management of system resources. This paper presents an application suitable algorithm for the generation of such control sequences on the basis of hierarchical Petri Nets.

1 Introduction

Rehabilitation robots were designed to support disabled persons in daily life situations as well as in the working environment. One category of rehabilitation robots are wheelchair mounted manipulators, like MANUS [1], RAPTOR [2] or the Weston Robot Arm [3]. These systems are flexible in use but offer services on a relatively low level of abstraction i.e. direct control of the manipulator. This puts a high cognitive load on their users, so that using these technical aids turns out to be an exhaustive activity [4]. To overcome this limitation systems like FRIEND [5], KARES I/II [6] or ISAC [7] have been introduced. These systems observe parts of the environment by means of sensors, so that some specific manipulation tasks can be performed autonomously (e.g. grasping an object by means of camera information [8]). Even though this relieves the user from the direct control of the manipulator, it is still necessary to reason about the sequence of operations during the execution of a higher abstraction level task (e.g. pouring or serving a drink).

In order to offer services on a higher level of abstraction, task planning becomes a necessary requisite. The system has to generate executable control sequences, in order to reach the goal related to the high abstraction level command. Domain independent planning approaches from the field of artificial intelligence are unsuitable for the use of real world robotic, especially rehabilitation robotic systems. This is due to the fact that the problem of task planning belongs to the class of NP-hard problems [9], so that for domain independent planning systems the required planning may exceed all tolerable time limits within real world robotic scenarios. An alternative approach is to use domain dependent planning systems, like in the field of assembly planning. Methods from the latter mentioned field make use of task representation data, like AND/OR-graphs or Petri Nets, that define the structure of the task to be solved in advance [10].

In [11] we introduced a task planning system for the rehabilitation robot FRIEND that uses assembly planning methods. The presented task planning component generates executable control sequences for the robotic system after the receipt of a command on a high level of abstraction. This paper concentrates on the generation of control sequences that take parallel executable operators into account. This aspect is important if the underlying control architecture supports the execution of parallel activities [12]. For instance, the system may possess more than one actuator (e.g. manipulator, wheelchair and pan-tilt heads), observation and manipulation processes have to run in parallel (e.g. for obstacle avoidance) or the user has to interact with the system during task execution. Generally spoken, the consideration of parallel activities enhances the exploitation of the system, shortens the required execution time and is inevitable for observation or interaction purposes.

The paper is organized as follows: After this introductive part chapter 2 starts with a short presentation of the planning system introduced in [11]. The hierarchical task representation used for the planning process is presented and the programming of the top level task representation with AND/OR-nets by an application programmer is explained. The resulting net is transformed automatically into an equivalent Petri Net, which is required by the succeeding planning process described in Chapter 3. Here, the planning problem on the basis of Petri Nets with parallel fireable transitions (i.e. the parallel executable operators) is discussed and an application suitable[1] algorithm is presented. Due to the used planning hierarchy the generated sequence of operations on this level of abstraction cannot be executed directly by the system. This entails a decomposition of operators which is described in chapter 4. The decomposition process takes care of resource management (i.e. sensors and actuators) in order to avoid possible blocking situations. Within the succeeding chapter 5 the consequential experimental results gained with a simulation system are presented and discussed. Finally, chapter 6 concludes the paper and gives an overview of future investigation and development steps.

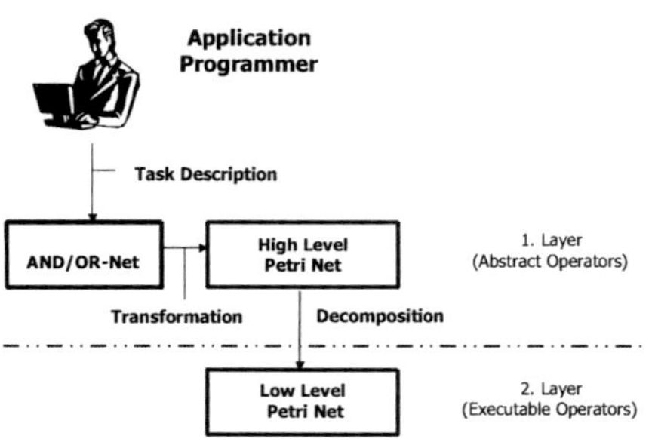

Fig. 1 Structure of Hierarchical Task Representation

2 Hierarchical task representation

With respect to the current state of art in rehabilitation robotics [13] high abstraction level command stands for the (semi-)autonomous processing of tasks like "pour a drink", "open a drawer" or "pick up objects from the floor". Even though a fully autonomous execution would be desirable, from the developers point of view it is more realistic to make use of the cognitive capabilities of the user. The

[1] Within this application domain *application suitable* is defined as: The system will be able to react to environmental changes or user interactions within a time period that is smaller than the average execution time of an autonomous operation of the system. Additionally, it should be possible to use the system ergonomically (i.e. too long reaction times are intolerable from the user's point of view).

main objective is to compromise about the two conflictive requirements: minimal user interaction and controllable complexity of the system. Our first solution for this problem can be found in [11]. The paper introduces a multi-layered control architecture with an embedded planning and execution component for the FRIEND system. The planning component makes use of a hierarchical task representation data structure that will be introduced in the following paragraph.

Fig. 1 depicts the general structure of the hierarchical task representation. On the top level an application programmer (e.g. care personnel) describes the task to be processed by means of so-called AND/OR-nets. AND/OR-nets, first introduced by Cao and Sanderson [10], are enhancements of task representation data structures from the field of assembly planning. These data structures are compact and efficient to search.

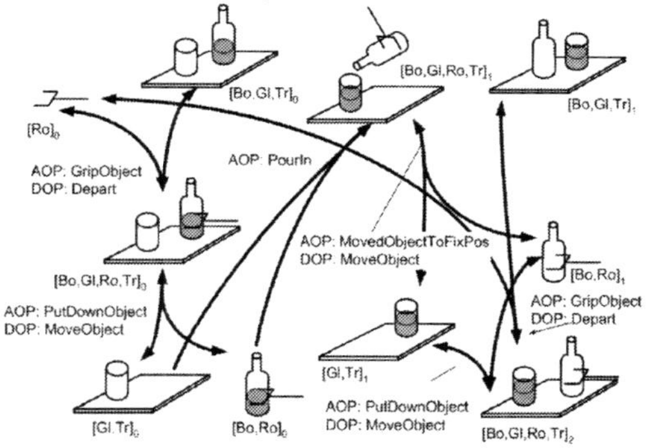

Fig. 2 AND/OR-Net corresponding to the "pour in a drink" task

The input of an AND/OR-net will be explained informally by means of the scenario depicted in fig. 2 (see [10] for a formal definition): The task to be described is "pour a drink". Therefore, a bottle (Bo) has to be gripped with the robot arm (Ro), moved to a glass (Gl) where the "pour-operation" is started. Afterwards, the bottle will be placed back on the tray (Tr). For simplification reasons it is assumed that an empty glass as well as an open and filled bottle are standing on the tray.

Fig. 2 depicts the resulting AND/OR-Net. Each node inside the net consists of one or more objects involved in the task scenario. By default, the composition of more than one object inside a node expresses that these objects are in physical contact. For example, $[Bo,Gl,Tr]_0$ represents the filled bottle as well as the empty glass standing on the tray. In order to state the situation more precisely predicates can be added to the nodes by the user. For instance, it might be reasonable to describe that the bottle is filled and the glass is empty. These predicates will be used for reasoning about the current state afterwards (see [10;14] for more details).

After the application programmer has defined all nodes that represent all possible (from his/her point of view) environmental and system situations (short: *situation*), they have to be connected with so-called AND-arcs. In fig. 2 it can be seen that each arc is connected with a single node on the one side and with two nodes on the other side. This expresses that an operator, like GripObject( ) or PutDownObject( ), will be applied on the nodes so that a new node emerges. It depends on the type of operator whether two nodes will be transformed into a single one or vice versa. For example, applying the operator GripObject( ) on the nodes $[Ro]_0$ (i.e. the gripper of the robot arm is empty) and $[Bo,Gl,Tr]_0$ produces the node $[Bo,Gl,Ro,Tr]_0$. The latter node stands for the situation that the gripper grasps the bottle, so that all object are in physical contact. Performing a Depart( ) operation afterwards separates the objects again. With respect to the field of assembly planning GraspObject( ) belongs to the type assembly operator (short: *AOP*) whereas Depart( ) is a representative for a disassembly operator (short: *DOP*).

After the connection of all nodes is completed the last step to be performed is to associate the high abstraction level command with a situation inside the AND/OR-net. In fig. 2, for example, the situation associated with the pour in command is: $[Bo,Gl,Tr]_0$ and $[Ro]_0$. It is evident that the problem of creating a feasible sequence of operations is now reduced to a graph search problem, in case that reasoning about the current situation can be assumed (see [11] for details).

Going back to the task representation depicted in fig. 1 the succeeding step after the AND/OR-net construction is the transformation into an equivalent Petri Net. The transformation process is described informally in fig.3. For a formal description see [10]. Although task planning on AND/OR-Net data structures is possible the transformation and postponed planning with Petri Nets offers certain advantages from the automation's point of view. Many software tools as well as sophisticated analysis algorithms are available, so that on a certain level of abstraction the control sequence generation process can be verified formally.

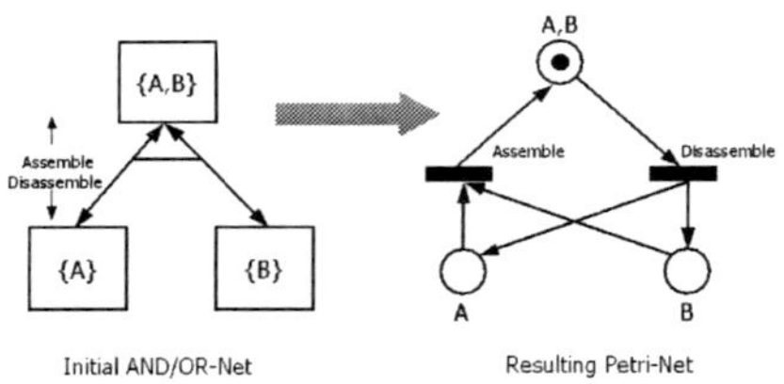

Fig. 3 Informal description of the AND/OR-Net to Petri Net transformation

The application suitable algorithm for the generation of control sequences on basis of the emerged Petri Net is described in the succeeding chapter. Within the current context it is important to know that the resulting control sequence cannot be executed by the system directly. The description of operators on this level of abstraction, like GripObject( ) depicted in fig.2, encapsulates the underlying technical system, so that a non-robotic-expert easily becomes able to input task knowledge without detailed information about manipulators, sensors or human-machine interfaces. All operators on this level of abstraction are composed of elementary operators, like a camera controlled fine approach process, that are autonomously executable by the system. For this reason, operators on AND/OR-level are termed as composed operators (short: *COP*) whereas the executable ones are named as elementary executable operators (short: *EEOP*).

The set of EEOPs comprises closed loop control operations with sensorial feedback, monitoring actions or man-machine interactions. For instance, the system offers a grasp object facility that continuously uses information from a gripper mounted camera. By means of this EEOP the gripper of the manipulator can be driven into the vicinity of an object to be gripped. In case the distance between the gripper and the desired object exceeds a predetermined limit the usage of a coarse approach EEOP that exploits environmental information of a stereo-camera system is preferable. Because it cannot be assumed that the system will be able to detect desired objects autonomously [13], the user will be included into the process of classification. This detection facility is encapsulated as EEOP as well.

A system programmer develops a COP by creating a Petri Net that uses EEOPs as transitions. From the system programmer's point of view, these nets have to consider all situations that might occur during the "execution" of a COP. The objective is to offer operators that are able to cope with dynamic environmental changes during task processing. By using Petri Nets the formal verification of this requirement becomes possible. For instance, the system programmer can verify that the Petri Net will reach a predetermined target marking from all possible initial markings. Because these markings are associated with target and initial situations this statement proofs that the operator will process its task correctly. It has to be mentioned that this kind of formal verification is restricted to the specifications of the system programmer.

Resuming the idea of task representation depicted in fig. 1, the necessity of operator decomposition from the abstract operator level down to the executable level is obvious now. After the creation of a control sequence on the highest level all involved operators have to be decomposed, so that eventually a sequence of EEOPs emerges. Beside the aspects of operator abstraction for task knowledge programming and encapsulation of system details the decomposition approach reduces the required control sequence generation time. The reason for the latter aspect is the nonlinear dependency between the control sequence generation time and the number of considered situations, as shown in the following chapter.

3 Application suitable generation of control sequences

So far, the structure of task representation on the basis of a Petri Net hierarchy has been introduced. In this paragraph an application suitable algorithm for the generation of control sequences that makes use of this data structure is presented. Due to the restricted size of the paper it will be assumed that the system "knows" the current environmental situation as well as its internal state. This information is used in both levels of the task representation for the determination of the initial situation required by the planning process. The description focuses on the generation of parallel executable operators. The inclusion of parallel activities enhances the exploitation of the system, shortens the required execution time and is inevitable for observation or interaction purposes. For the sake of readability all necessary definitions related to Petri Nets and graphs have been centralized in the Appendix.

For the control sequence generation process it is assumed that each place of a marked Petri Net PN_m (see def. A1, A2) is associated with a single element of a situation interesting for the description of the task. For instance, $p_i \in P$ stands for the instantiated predicate standsOn(Bottle,Tray). Additionally, each transition is associated with an operator offered by the system (i.e. COP or EEOP). Metaphorically spoken, firing a transition under a current marking stands for the execution of the transition associated operator in the current situation (, see def. A3, A4). Then, the problem of control sequence generation is equivalent to the problem of finding a finite firing sequence of transitions $\sigma = t_1, t_2, t_3, \ldots ,t_n$ leading from the initial marking M_0 to the target marking M_t (short: $M_0[\sigma > M_t)$. It is assumed that the initial marking is associated with the current situation whereas the target marking describes the desired situation associated with the high abstraction level command. Therefore, finding an appropriate control sequence $M_0[\sigma > M_t$ is equivalent to search for a path within the reachability graph RG (see def. A5, A6) between the nodes M_0 and M_t. The problem of control sequence generation has been reduced to a graph search problem wherefore well known algorithms are available. The next question to be answered is how to find control sequences that take concurrent activities into account.

Cao and Sanderson proofed that Petri Nets resulting from the AND/OR-net transformation belong to the class of Condition/Event-Systems *CES* [10], see def. A7. For the latter mentioned systems the question of whether operators might be executed concurrently or sequentially can be answered by means of the so-called diamond property[2] [16], see def. A8. The diamond property expresses that two or more concurrently fireable transitions within a CES don't possess an intersection in their pre- and post-places and they are enabled under the current marking. A result of this property is the appearance of the so-called concurrency pattern (see def. A10) within two different firing sequences that produce the same marking but differ in their transition firing order.

Our first approach for a control sequence generation algorithm with parallel operators makes use of the concurrency pattern [17]. The suggested algorithm uses a fully expanded reachability graph RG associated to a CES resulting from an AND/OR-net transformation. On the basis of this RG all feasible sequential firing sequences from the initial to the target marking are extracted by means of a depth-first search algorithm. Due to the finite size of the RG and the integrated loop avoidance strategy (i.e. it is not allowed to reach the same marking more than one time within a firing sequence) it has been proven that the extraction algorithm stops after a finite processing time. A succeeding investigation step computes all causal independent operators by exploiting the concurrency pattern. Each time such a pattern appears in two paths these paths are merged by means of replacing the equivalent step (see def. A9) at the position of the pattern. At last the "best" plan is selected taking the optimisation criteria "minimum weight of the whole plan" into account. Currently, the weight for a single transition or step is set to one, so that the shortest path will be judged as optimal. Future investigations concentrate on the problem of choosing appropriate weights for each transition that express its efforts (e.g. computational, sensorial) and robustness (i.e. success for execution).

The algorithm suggested in [17] has some drawbacks that makes it unsuitable for applications in the domain of rehabilitation robotic systems. First, the algorithm doesn't eliminate "unreasonable" paths in advance. The anthropomorphic term "unreasonable" shall be clarified by means of the scenario depicted in Fig. 4. Within this scenario two manipulators have to perform two causal independent tasks. The left manipulator (Robot 1) has to pour in a drink whereas the right one (Robot 2) shall move a glass from the right to the left table. The suggested algorithm might create a path that consists of a partial operation sequence like GripObject(Robot 2, Glass 2), Depart(Robot 2), GripObject(Robot 2, Glass 2) etc.. Robot 2 is caught in a cycle of reversible operations. For a human observer it is evident to judge the execution of these actions as "unreasonable". But for the suggested algorithm these paths will be taken into account as long as the insertion of different operations of Robot 1 will lead to distinguishable markings within the reachability graph.

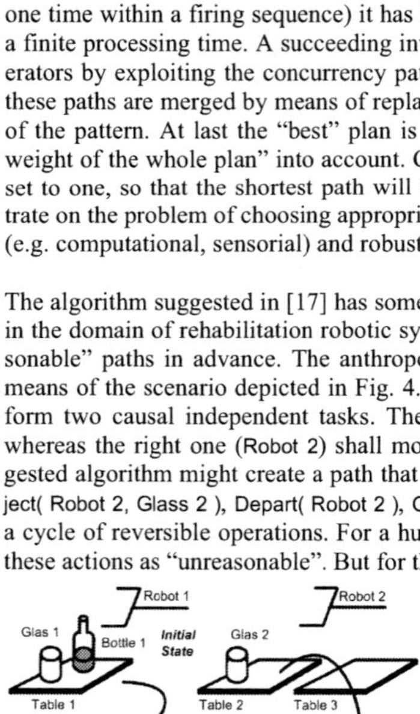

Fig. 4 Example for causal independent tasks

Even though these "unreasonable" paths will be eliminated within the final "best" plan selection process, the accompanied calculation complexity as well as the required memory space avoid its application[3]. Due to these drawbacks an algorithm that exploits concurrency information before the selection of the "best" plan is suggested in this paper. The idea is to construct a reachability

[2] This property is named „diamond property" because within the graphical representation of the Petri Net related reachability graph the pattern of the arcs associated with the concurrently fireable transitions can be interpreted as a diamond.

[3] The number of possible paths N_P is exponentially dependent on the number of parallel activities. The succeeding concurrency check has a quadratic dependency on the product of N_P and the average path length.

graph on the basis of transitions and steps in order to avoid the complexity for the calculation of all feasible paths. For this type of modified *RG* it is sufficient to search the shortest path from the initial to the target marking, because the resulting path already considers concurrent activities. This path can be judged as optimal with respect to the above mentioned optimisation criteria.

The difference to the construction of a reachability graph described in algorithm A.1 is the fact that also all enabled steps will be taken into account for the calculation of successor markings. Therefore, the algorithm has to cope with the complexity of step calculation. The first step for the determination of all enabled steps is to determine the set of all enabled transitions ($T_e \subseteq T$) (see def. A3). According to the diamond property, permutations of these transitions may form enabled steps. We have to introduce the following definition for the formulation of an efficient step verification algorithm.

Definition 1: Maximum set of concurrent transitions (MSCT)
Let $T_e \subseteq T$ be the set all enabled transitions under a marking M_c in a marked Petri Net. A maximum set of concurrent transitions $T_{Cmax} \subseteq T_e$ consists of transitions that are all concurrent to each other i.e. $\forall t_i, t_j \in T_{C\max} \Rightarrow t_i \parallel t_j$ and no set T^*_{Cmax} with $T_{Cmax} \subset T^*_{Cmax} \subseteq T_e$ exists that has the same characteristic.

A MSCT defines the set of transitions from which each subset with at least two elements forms an enabled step. Fig. 5 visualizes the concurrency relation by means of a concurrency graph G_c. Each node of G_c represents a transition from T_e. Two nodes t_i, t_j are connected by an undirected arc if $t_i \parallel t_j$. Examples of MSCTs in fig. 5 are $\{t_1,t_3\}$ or $\{t_1,t_2,t_4\}$. The latter MSCT $\{t_1,t_2,t_4\}$ produces the set of enabled steps $\{(t_1,t_2,t_4),(t_1,t_2),(t_2,t_4),(t_1,t_4)\}$. Because of the intransitivity[4] of the concurrency relation the intersection of two different MSCT is not empty. From fig. 5 it can be seen that the problem of finding all MSCTs under a marking is equivalent to the problem of finding all cliques (see def. A11, A12) in G_c.

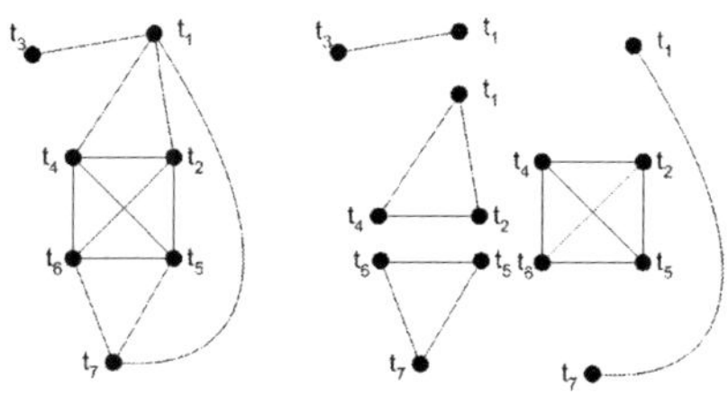

Fig. 5 Concurrency graph (left) and its cliques (right)

An efficient algorithm known for generating all cliques of a graph can be found in [18]. Even though there is no known exact analysis of its running time, it can be said that it cannot be expressed as a polynomial in $|V|$ and $|E|$. Generally, this would restrict its technical applicability for concurrency analysis purposes of marked Petri-Nets. But for the application domain treated in this paper the algorithm is suitable (see chapter 5 for an example). Here, the value of $|V|$ expresses the number of possible alternatives or concurrently executable operations in a certain situation. This value is restricted by the number of independent actuators, sensors and pre-programmed planning alternatives.

As mentioned above, by means of the acquired MSCTs all distinguishable steps under the current marking M_c can be determined, so that the construction of the enhanced reachability

---

[4] $\neg(\forall t_i, t_j, t_k \in T_e : (t_i \parallel t_j) \wedge (t_j \parallel t_k) \Rightarrow (t_i \parallel t_k))$

graph becomes possible. On basis of this reachability graph the path with minimum weight from the initial marking to the target market can be found by means of Dijkstra's algorithm. In the worst case this algorithm has a quadratic complexity related to the number of nodes. Due to "Bellman's Principle of Optimality" it is possible to find the shortest path without the investigation of all nodes of the graph [19]. This property is important for this application domain, because, as already explained in the example of fig. 4, the reachability graph might contain "unreasonable" paths and situations. Therefore, a partial construction of the reachability graph interleaved with Dijkstra's algorithm, as suggested in [10] for sequential executing robotic systems, is used for this application.

4 Operator decomposition

After a control sequence on the AND/OR-net level has been generated by means of the presented algorithm, each COP of this sequence has to be decomposed, so that an executable sequence of EEOPs emerges. The decomposition process creates a control sequence on basis of COP related Petri Nets. Because the structure of these Petri Nets differs from the AND/OR-net related ones, some modifications of the control sequence generation algorithm are necessary.

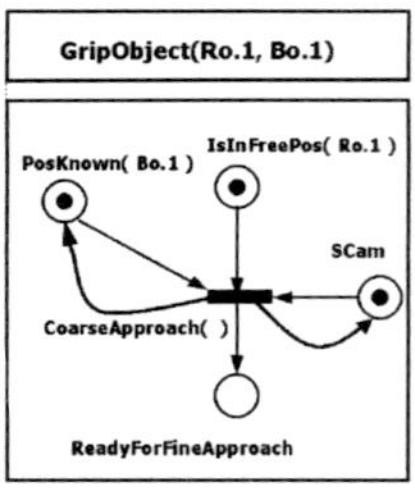

Fig. 6 COP GripObject

Fig.6 depicts an example of a COP related Petri Net. For the sake of simplification only a small part of the real COP implementation is depicted. The Petri Net possesses the transition (EEOP) CoarseApproach( ) with the pre-conditions robot (Ro.1) is in free position, its position is known and a stereo-camera system (SCam) is available. In case that all conditions are fulfilled the CoarseApproach( ) action can be started. After its execution, the system is ready for the succeeding fine approach process.

In contradiction to AND/OR-net related Petri Nets this type of Petri Nets contains loops, like the check for camera availability. The introduction of loops for the design of these Petri Nets offers different advantages: First, it reduces the complexity of the net. Second, it reduces the size of the Petri Net related reachability graph. At last, no unnecessary concurrently enabled transitions have to be considered during the control sequence generation. A negative effect is that these Petri Nets don't belong to the class of CES. This entails that the diamond property cannot be used for the concurrency check. Because these nets still possess the semantics of CES, the computational effort of this check is still low in comparison to the concurrency check in generally marked Petri Nets. We introduce the following definition for a concurrency check suitable for COP related Petri Nets.

Definition 2: Concurrency verification for COP related marked Petri Nets
Let $t_i, t_j \in T, t_i \neq t_j$ be two arbitrary transitions of COP related marked Petri Net and M_C its current marking: Iff $({}^\bullet t_i \cup t_i^\bullet) \cap ({}^\bullet t_j \cup t_j^\bullet) = \{\} \wedge (\forall p \in ({}^\bullet t_i \cup {}^\bullet t_j) : M_C(p) = 1) \wedge$ $(\forall p \in (t_i^\bullet \cup t_j^\bullet) : M_C(p) \leq 1)$ then $t_i \parallel t_j$.

The next problem to be solved during the decomposition process is the problem of resource administration. Imagine a control sequence for the robotic system like in fig. 4 that possesses two GripObject( ) operations to be executed in parallel. In case only a single stereo-camera system is available, a resource conflict appears, as shown in the following picture (see fig. 4 left).

Both operations require the same system resource, so that the associated nets in addition with their initial and target marking have to be merged before the plan generation can be started (see fig.4 right). After merging, the plan generation algorithm for the AND/OR-net layer can be used, if the concurrency check modification (see def. 2) is taken into account. It has to be mentioned that checking for conflicts that may arise due to shared system resources is sufficient. Possible resource conflicts due to shared actuators or manipulated objects have been solved already on the AND/OR-net level. In order to verify whether a resource conflict might be possible or not, each COP related Petri Net is connected with a list of required system resources.

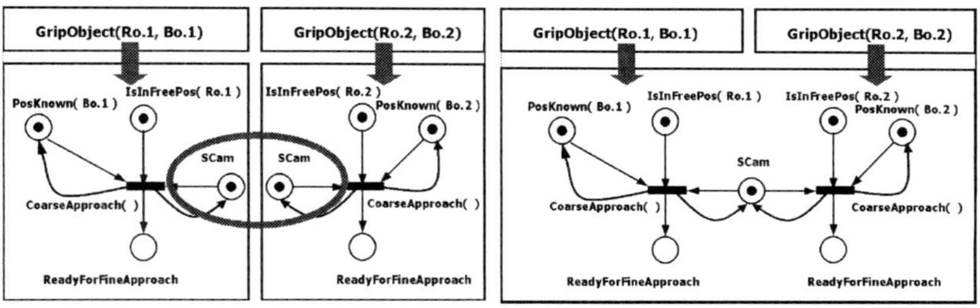

Fig.7 Petri Nets before(left) and after(right) the merging process

5 Experimental results

The developed algorithm has been integrated into the planning program presented in [11]. As benchmark the scenario depicted in fig. 4 was chosen, due to its high degree of concurrent activities. Additionally, it is representative for the complexity of interesting high abstraction level tasks. For reasons of authenticity COPs and EEOPs related to the FRIEND system were used for operator modelling. The planning program was executed on a 1.2 GHz Pentium III processor with 256 MB RAM. This configuration equals computer hardware available in current rehabilitation robotic systems. The execution time and intermediate results were compared with the algorithm presented in [17] and a Dijkstra algorithm on the basis of a reachability graph without step consideration. The latter algorithm also makes use of the interleaved searching and graph construction method. The comparison results are listed in the following table.

	D_S Dijkstra with steps	**D_T** Dijkstra without steps	**BF** [17]
Execution time [ms]	600	6	Abortion after 10 min[5]
Nodes in RG	34	34	40
Number of operations on COP level	6	9	n/a
Number of operations on EEOP level	25	26	n/a

[5] At abortion time more than 1000 possible paths were generated.

Table 1Execution results of control sequence generation algorithms

It can be seen that the calculation time of D_S is approx. 100 times higher than of D_T . Reasons for this are the calculation of steps and possible merging of nets. Even so D_S can be rated as suitable for the proposed application domain where the average time for the execution of an EEOP is approx. 1-2 seconds. Additionally, in real world scenarios planning and execution are performed in an interleaved manner. This means that plans won't be created for a scenario in a whole but stepwise on different levels of abstraction. Additionally, this reduces the required time for plan creation.

In table 1 is interesting to see the effect that even though the COP sequence of D_S is shorter than for D_T (due to concurrently executable operators) the length on EEOP level is almost equivalent. This is due to the assumption that the modelled system possesses a single stereo-camera system. The resource management process during operator decomposition creates sequences of operations even though associated COPs have been estimated as concurrently executable. This aspect should be taken into consideration for the integration in concrete systems, in order not to waste calculation time. For example, in case of FRIEND the application of D_S on COP level is unreasonable, because the system possesses a single manipulator.

6 Conclusions

In rehabilitation robotic systems it is desirable to offer high abstraction commands. This however entails the problem of task planning. This paper presents a control sequence generation algorithm that improves our task planning approach suggested in [11]. The algorithm takes concurrent activities into account, so that the utilisation of the system is enhanced. As shown, within this context the problem of high abstraction level command control is equivalent to search a firing sequence from an initial to a target marking in a marked Petri-Net. In case both markings are given the search process can be performed on the reachability graph associated with the Petri-Net. In order to take parallel executable operations into account, the construction of the reachability graph uses transitions as well as steps. It is shown, that, even though the consideration of steps (and of parallel executable operations in general) entails higher computational effort, the plan generation process is still suitable within the domain of rehabilitation robots. Additionally, the suggested plan generation algorithm is usable on both task representation levels, if small modifications of the algorithm are taken into account. On the basis of the modified reachability graph the best "path" is found by Dijkstra's algorithm. This algorithm searches for a path with minimum weight. This aspect is subject of future investigations. Currently, transitions and steps possess equal weight, so that "best" is equivalent to shortest. For future enhancements these weights have to express characteristics of the associated operations, like robustness in execution, necessary computational effort, probability of success etc. Additionally, the weights of the COP associated transitions should be influenced by the consumption of system resources, so that their impossible concurrent activation can be avoided on the AND/OR-net level directly.

Appendix

In the following all necessary definitions of Petri Nets (see [15,16]), Petri Net related algorithms and graphs (see [18]) are given.

Definition A1: A *Petri Net* is a tuple $PN = (P, T, F, C, W)$ where
(P, T, F) is a net with:

P:	Non-empty and finite set of places	
T:	Non-empty and finite set of transitions	
F:	Flow relation with $F \subseteq (P \times T) \cup (T \times P)$	

C: $P \to \{1,2,3,\ldots\} \cup \{\infty\}$ – partial capacity restriction (default ∞)

W: $F \to \{1,2,3,\ldots\}$ – weight function (default 1)

Definition A2: A *marked Petri Net* is a pair $PN_m = (PN, M_0)$ where

PN: Petri Net

M_0: $P \to \{0,1,2,3,\ldots\}$ – a marking (here initial marking) satisfying:

$$\forall p \in P : C(p) = \infty \vee M_0(p) \le C(p)$$

Definition A3: *Enabled transition t* under a marking M_C in a marked Petri Net

If for a transition $t \in T$ the set ${}^\bullet t = \{p \in F(p,t)\}$ denotes its pre- and $t^\bullet = \{p \in F(t,p)\}$ its post-conditions then t is said to be enabled under the current marking M_C if $\forall p \in {}^\bullet t : M_C(p) \ge W(F(p,t))$ and $\forall p \in t^\bullet : M_C(p) + W(F(t,p)) \le K(p)$.

Definition A4: *Firing of an enabled transition t* under a marking M_C in a marked Petri Net

An enabled transition is said to be fired if the current marking M_C is transformed into the successor marking M_C' (short: $M_C[t > M_C')$ in the following manner:

$$M_C'(p) = \begin{cases} M_C(p) - W(F(p,t)) & \text{if } p \in {}^\bullet t \setminus t^\bullet \\ M_C(p) + W(F(p,t)) & \text{if } p \in t^\bullet \setminus {}^\bullet t \\ M_C(p) - W(F(p,t)) + W(F(p,t)) & \text{if } p \in t^\bullet \cap {}^\bullet t \\ M_C(p) & \text{otherwise} \end{cases}$$

Definition A5: *The reachability set* of a marked Petri Net (PN, M_0) is the set $R(PN, M_0)$ such that $(M \in R(PN, M_0)) \Leftrightarrow (\exists \sigma : M_0[\sigma > M)$ where σ stands for a sequence of fired transitions.

Definition A6: The *reachability graph RG* of a marked Petri Net PN_m is a directed graph for which the nodes are all markings which are elements of $R(PN, M_0)$ and there is an oriented arc from marking M to marking M' iff $\exists t \in T$ such that $M[t > M'$.

Algorithm A1: *Construction of a reachability graph*

Input: Marked PN_m

Output: Reachability graph $RG = (V,E)$

Initialize: $ListM := \{M_0\}$, $V := \{M_0\}$, $E = \{\}$

while $ListM \neq \{\}$ **do**

  take an arbitrary $M \in ListM$; $ListM := ListM \setminus \{M\}$

  for $\forall t \in T_e$ (i.e. enabled transitions under M in PN_m)

    $M' := M[t{>}M'$; $E := E + \{M[t{>}M'\}$

    if $M' \notin V$ **then** $V := V + \{M'\}$, $ListM := ListM + \{M'\}$

  endfor

endwhile

Definition A7: A *Condition/Event-System CES* is a marked Petri Net with the following restrictions: $K : S \to \{1\}$; $W : F \to \{1\}$; $M_0 : S \to \{0,1\}$ and $x^\bullet \cap {}^\bullet x = \{\}$ (i.e. *CES* are loop free).

Definition A8: *Diamond property* for *CES*

Let $t_i, t_j \in T, t_i \neq t_j$ be two arbitrary transitions of a *CES* and M_C its current marking. Iff

$$({}^\bullet t_i \cup t_i^\bullet) \cap ({}^\bullet t_j \cup t_j^\bullet) = \{\} \wedge (\forall p \in ({}^\bullet t_i \cup {}^\bullet t_j) : M_C(p) = 1) \wedge (\forall p \in (t_i^\bullet \cup t_j^\bullet) : M_C(p) = 0)$$

then $t_i \parallel t_j$ i.e. t_i and t_j can be fired concurrently (i.e. they are causal independent).

Definition A9: *Step*
Two or more concurrently enabled and fired transitions are called a step.

Definition A10: *Concurrency pattern* in firing sequences of *CES*
Two arbitrary transitions t_i and t_j of an arbitrary *CES* are concurrently fireable under a marking $M_k \Rightarrow M_k[t_i, t_j > M_l$, $M_k[t_j, t_i > M_l$ and $M_k[t_j \parallel t_i > M_l$ are feasible firing sequences.

Definition A11: *Complete subgraph* of an undirected graph $G = (V,E)$
Let V bet the set of vertices and E the set of edges of the graph G. $G' = (V',E')$ is called a complete subgraph of G iff $V' \subseteq V$, $E' \subseteq E$ and there is an edge in E' between every pair of vertices in V'.

Definition A12: *Clique* of a graph G
A maximal complete subgraph of a graph is called a clique; that is, a clique of G is a subset of the vertices of G such that there is an edge in G between every pair of vertices in the subset and, furthermore, such that the subset is not contained in any larger subset having the same property.

7 References

[1] Dallaway, J. L.; Jackson, R. D.; Timmer, P.H.A. "Rehabilitation Robotics in Europe", IEEE Transactions on Rehabilitation Engineering 1995, Vol. 3, pp. 33-45

[2] Mahoney; R. M. "The Raptor Wheelchair Robot System", in Mokhtari, M. (Ed.) Integration of Assistive Technology in the Information Age, proceedings of ICORR 2001, Evry, France

[3] Hilman, M. R. et. al. "Design of a Wheelchair Mounted Robot", Proceedings of AAATE 1999, November 1-4, 1999, Düsseldorf, pp. 316-321

[4] Bagchi, S.; Biswas, G.; Kawamura, K., "Task planning under uncertainty using a spreading activation network", in Systems, Man and Cybernetics, Part A, IEEE Transactions on , Volume: 30 Issue: 6 , Nov. 2000

[5] Martens, C.; Lang, O.; Ruchel, N.; Ivlev, O.; Gräser, A. "A FRIEND for Assisting Handicapped People" March 2001 issue of IEEE Robotics and Automation Magazine

[6] Z. Bien et. al. "A Wheelchair Robot System and its Various Interface Methods for the Disabled Persons", The 1st Workshop on Technical Challenge for Dependable Robots in Human Environments, Seoul, Korea, May 21-22, 2001.

[7] Kawamura, K. et. al. "ISAC: foundations in human-humanoid interaction" Intelligent Systems, IEEE, Volume: 15 Issue: 4 , Jul/Aug 2000 Page(s): 38 –45

[8] Lang, O.; Martens, C; Gräser, A. "Realisation of a Semiautonomous Grippingskill for the Support of Disabled People" Robotik 2000 29./30. Juni 2000 Berlin, VDI Berichte 1552

[9] Dean, T.; Kambhampati, S. "Planning and Scheduling" in the CRC Handbook of Computer Science and Engineering, A. B. Tucker (Ed.), CRC press, 1997. pp614-636.

[10] Cao, T.; Sanderson, A. C. „Intelligent Task Planning Using Fuzzy Petri Nets" World Scientific Publishing 1996 ISBN 9810225563

[11] Martens, C.; Gräser, "Design and Implementation of a Discrete Event Controller for High-level Command Control of Rehabilitation Robotic Systems", proceedings of IROS 2002, September 30 - October 4, 2002, Lausanne, Switzerland

[12] Martens, C.; Kim, D.J.; Han, J.S.; Gräser, A.; Bien, Z. "Concept for a Modified Hybrid Multi-Layer Control-Architecture for Rehabilitation Robots" 3rd International Workshop on Human-friendly Robotic Systems, Taejon, Korea, January 21-22, 2002

[13] Dario, P. et. al. "EURON Research Roadmaps 2002" Research Roadmaps of the European Robotics Research Network 2002, http://www.euron.org

[14] Martens, C.; Schüttler, J.; Gräser, A. Logical Verification of AND/OR-net Structures for Task-Knowledge Representation in Service Robotic Scenarios" in Proceedings of the ICORR 2003, 23-25 April 2003, Taejon, Korea, http://icorr2003.rehabrobotics.org

[15] Baumgarten, B. "Petri Netze – Grundlagen und Anwendungen" BI Wissenschaftsverlag, Mannheim, 1990 ISBN 341114291X

[16] Rozenberg, G. „Elementary Net Systems“ in Petri Nets 2000, 21st International Conference on Application and Theory of Petri Nets, Introductory Tutorial, June 2000
[17] Kühn, J.; Märkl, S.; Schröder, C. „Handlungsplanung auf der Basis von Petri Netzen“ Project Thesis, Institute of Automation, University of Bremen, August 2001
[18] Reingold, E. M. et. Al. „Combinatorial Algorithms – Theory and Practice“, Prentice Hall 1977
[19] Aho, A. V.; Ullman, J. D. "Informatik – Datenstrukturen und Konzepte der Abstraktion" Thomson Publishing 1996, ISBN 3-8266-0242-0

Object grasping and obstacle avoidance under visual control of adjustable zoom-cameras

Olena Radchenko, Andreas Pape and Axel Gräser
Institute of Automation, University of Bremen, 28359 Bremen
radchenko@iat.uni-bremen.de, apape@iat.uni-bremen.de
and ag@iat.uni-bremen.de

This paper presents the visual servoing algorithm with adjustable zoom-cameras and a method of obstacle avoidance during object grasping and manipulation. Image-based visual servoing needs image features of the real scene. Here the method of virtual points with adjustable zoom-cameras, i.e. points without detectable image feature in the real scene is used. Pan-tilt and zoom control algorithms, integration of accordant control loops into the main system and its influence on a visual guided system will be discussed.

1 Introduction

The rehabilitation system FRIEND (figure 1) is developed at the IAT to be used by disabled people. The most important task for the system is to grasp and manipulate objects in the workspace of the system. The movements of the robot are controlled by stereo cameras using visual servoing. This is the well-known method realised in many applications worldwide. But most of these realisations are based on fixed cameras with constant focal lengths. The use of fixed cameras restricts the workspace of the system. But it is very important for service robots like FRIEND to have the workspace as big as possible. To prevent the situations when the robot arm leaves the common field of view zoom cameras mounted on pan-tilt-heads are used.

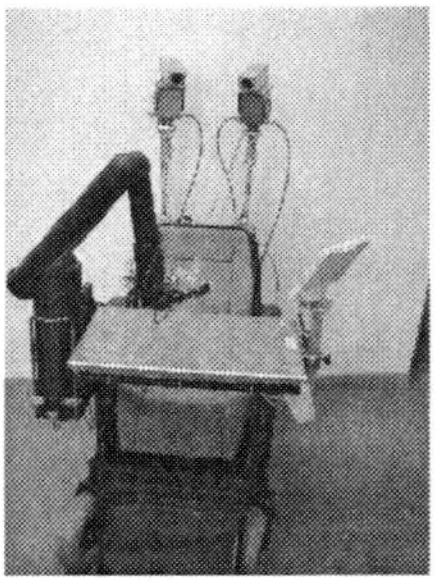

Figure 1: System FRIEND

Another important task is obstacle avoidance. If no obstacles are in the working area of the robot arm the visual servoing can be done directly. But usually in the workspace some obstacles must be avoided.
This paper is structured as follows: After the description of the workspace of the system the conventional visual control of a robot will be presented first. This explanation is followed by a description of an extended visual servoing method and an extended concept of virtual points to avoid the obstacles during execution of the task. The paper concludes with results of automatically gripping scenario.

2 Visual control of robot

2.1 Workspace of the system

Rehabilitation robots like FRIEND equipped with a robot arm are designed to support totally paralysed people in daily life tasks. A large workspace of the robot is essential. In the case of visual servoing of the robot this is determined by a combination of the robot arm workspace and the field of view of the stereo cameras system. Using fixed cameras with constant focal length the common field of view is restricted to a comparatively small area formed by the intersection of the common fields of view of the single cameras (figure 2, a). The common field of view must be optimised for a part of the workspace, e.g. for the tray mounted in front of the user. To enhance the usability, the possible tasks must not be restricted to handle objects on the tray. Even in the "gripping from the tray" scenario it often happens that the robot gripper during the execution of different tasks leaves the common field of view. Then the use of visual servoing methods is not possible any more. There are two ways to solve this problem: use more than one camera system for different applications or use one system with adjustable cameras. We have chosen the second one. Both cameras are mounted on pan-tilt heads and have variable focal length (figure 2, c). In this case the common field of view covers the whole workspace of the robot arm (figure 2, b).

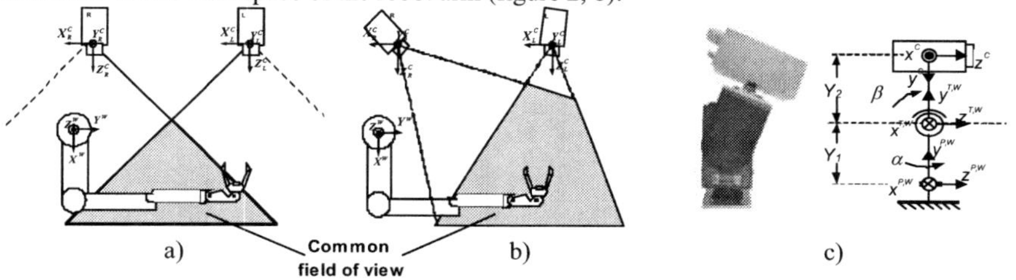

Figure 2: Coordinates of the system and field of view of the cameras

2.2 Visual servoing with fixed cameras

2.2.1 Coordinate systems and nomenclature

The mapping from the robot motion to the visual motions includes a hand-eye coordinate transformation and the camera perspective map. In figure 2 the coordinate systems, which are used for the transformations, are depicted.
We employ the following nomenclature:

Symbol	Meaning
W	robot based coordinate system
C_i	position of camera i relative to W i = l, r
R_i	rotation matrix
r^W	cartesian coordinates relative to W of an observable reference point of a robot
$r^I = [u_i, v_i]$	image coordinates of observable reference point of a robot
u^W	robot control input
g	mapping from Cartesian coordinates to stereo image coordinates
P and T	Pan and Tilt coordinate systems accordantly (left and right)

Table 1: nomenclature

The camera positions are defined by the optical centre of the camera lens. The field of view of the cameras is defined as cone and a common field of view of two cameras is the intersection of the fields of view of left and right cameras (figure 2, b).

2.2.2 Conventional visual controller

The problem of visual feedback control is defined as generating of cartesian motion from image observations. The conventional visual controller (figure 3) uses the location of features on the image plane directly for feedback. The controller includes Image Jacobian matrix, which describes the relationship between changes in the image plane and changes in the world coordinate.

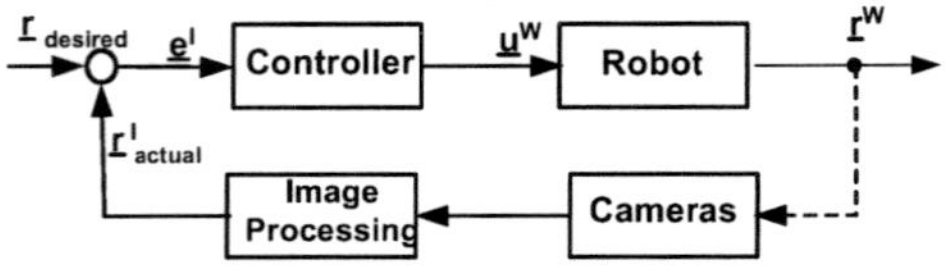

Figure 3: Conventional visual controller

The mapping from the robot motion to the visual motions includes a hand-eye coordinate transformation and the camera perspective map. In classical approach both cameras are assumed to have identical constant focal length [1]. Projection of the object in the workspace into the camera coordinate system can be described by a transform matrix (1). This matrix contains the information about relative position and orientation of systems:

$$
{}_{C}\underline{T}^{W} = \begin{bmatrix} rot_{11} & rot_{12} & rot_{13} & 0 \\ rot_{21} & rot_{22} & rot_{23} & 0 \\ rot_{31} & rot_{32} & rot_{33} & 0 \\ 0 & 0 & 0 & 1 \end{bmatrix} \begin{bmatrix} 1 & 0 & 0 & -C_X^W \\ 0 & 1 & 0 & -C_Y^W \\ 0 & 0 & 1 & -C_Z^W \\ 0 & 0 & 0 & 1 \end{bmatrix}
= \begin{bmatrix} rot_{11} & rot_{12} & rot_{13} & -C_X^W rot_{11} - C_Y^W rot_{12} - C_Z^W rot_{13} \\ rot_{21} & rot_{22} & rot_{23} & -C_X^W rot_{21} - C_Y^W rot_{22} - C_Z^W rot_{23} \\ rot_{31} & rot_{32} & rot_{33} & -C_X^W rot_{31} - C_Y^W rot_{32} - C_Z^W rot_{33} \\ 0 & 0 & 0 & 1 \end{bmatrix} \quad (1)
$$

Then adding stereo perspective projection (with pin-hole camera model) the observation of a point *r* will be given by:

$$
r^i = \begin{bmatrix} u_L \\ v_L \\ u_R \\ v_R \end{bmatrix} = T \begin{bmatrix} x \\ y \\ z \end{bmatrix} = g(r) \quad (2)
$$

Assuming that the image error $e = r^i - r^{i*}$ we define according to [1]:

$$
\begin{aligned} \dot{e}^I &= J_g(r) u^W \\ \dot{r} &= u^W \end{aligned} \quad (3)
$$

where $J_g(r)$ is the Jacobian of g at *r*.
In case of static mounted cameras the Jacobian matrix depends only on the location of the reference point (robot) in the world coordinate system $\underline{J} = \underline{J}(r)$ because of constant value of other components. It has been assumed that the robot reference position is known. But this information cannot be received directly and should be estimated.

2.3 Visual servoing with adjustable cameras

This is the extension of the visual servoing method, designed for using with an adjustable camera system. Because of additional degrees of freedom provided by pan-tilt heads the transformation between world and image coordinates have to be adapted. In this case we should take into account variable focal length and orientation of the cameras. Mathematically it means that a transform matrix

$$ {}_{C}T^{W} = \begin{pmatrix} -\cos(\alpha) & 0 & -\sin(\alpha) & 0 \\ -\sin(\alpha)\cdot\sin(\beta) & -\cos(\beta) & \sin(\beta)\cdot\cos(\alpha) & -\cos(\beta)\cdot Y_1 - Y_2 \\ -\sin(\alpha)\cdot\cos(\beta) & \sin(\beta) & \cos(\alpha)\cdot\cos(\beta) & \sin(\beta)\cdot Y_1 \\ 0 & 0 & 0 & 1 \end{pmatrix} \quad (4) $$

must be included in the transformation. As a result we will get a Jacobian matrix, which depends on the six additional variable parameters: two orientation angles (pan and tilt angle) and focal lengths for each camera:

$$ \underline{J} = \underline{J}(\underline{r}, \underline{\alpha}, \underline{f}) = \underline{J}\left(\underline{r}, \begin{pmatrix} \alpha_L \\ \beta_L \\ \alpha_R \\ \beta_R \end{pmatrix}, \begin{pmatrix} f_L \\ f_R \end{pmatrix} \right) \quad (5) $$

To control the robot with adjustable cameras the extended Jacobian matrix must be included in the control loop. For the control of the camera rotation and the focal length, two additional control loops are included (figure 4).

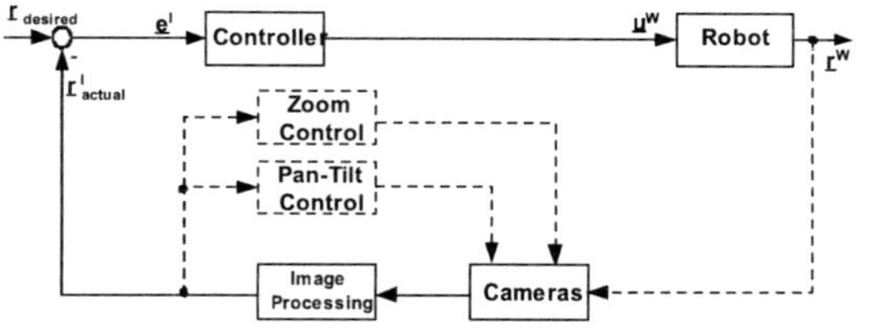

Figure 4: Extended visual controller

2.3.1 Pan-tilt head control

The main task of this control loop is the tracking of the marked object. Orientation of the camera is controlled, so that the tag always remains in the centre of the image. To achieve this the camera is actuated by a Pan-Tilt device. This Pan-Tilt allows the rotation of the camera around two axes (horizontal – pan axis – yaw and vertical – tilt axes – pitch)(Fig. 2, c).
We use the same basic idea as for robot control: we calculate an image error and use it to generate the set points for pan-tilts. The desired position is the midpoint of the image. The actual position is the midpoint between the robot gripper and the object. In this case we will get the image Jacobian matrix in form:

$$ J_{PTH} = \begin{bmatrix} f_L\varphi_{11}(\alpha_L,\beta_L) & f_L\varphi_{12}(\alpha_L,\beta_L) & 0 & 0 \\ f_L\varphi_{21}(\alpha_L,\beta_L) & f_L\varphi_{22}(\alpha_L,\beta_L) & 0 & 0 \\ 0 & 0 & f_R\varphi_{44}(\alpha_R,\beta_R) & f_R\varphi_{45}(\alpha_R,\beta_R) \\ 0 & 0 & f_R\varphi_{54}(\alpha_R,\beta_R) & f_R\varphi_{55}(\alpha_R,\beta_R) \end{bmatrix} \quad (6) $$

As expected left and right cameras don't have any influence on each other and can be controlled separately. The control loop for one camera in detail is presented on the figure 5, b.

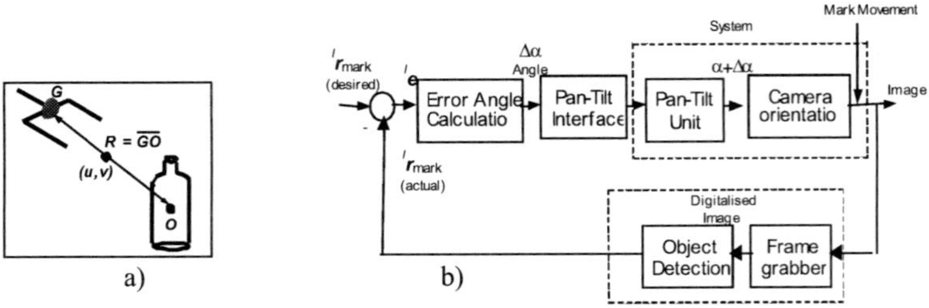

Figure 5: Visual servoing task and Pan-tilt control loop

During operation of the control loop we assume that the focal length of the cameras is adjusted so that objects of interest are visible in sufficient resolution and besides that the focal length of cameras is known.

2.3.2 Zoom control

Zoom lenses fit very naturally into the active vision paradigm since they provide a powerful degree of freedom for adjusting the appearance of the image. Depending on the task of the system zoom control might have different tasks:

- it might be advantageous to zoom out, yielding a broad overview of a large area in front of the camera (for the analysis of the work space of the system);
- controlling a zoom lens allows an adjustment of the image enabling a close look at a region or object of particular interest (for example, if the object on the tray is not identified because of too small size);
- it can preserve the "size" of an object in the image as it moves, creating more stable pictures (the zoom setting of the camera is then changed so that the object size fluctuation will be compensated).

However, the integration of zoom into a visual system can not be done without difficulty. The changes of the focal length must be known if any measurements are to be made. These parameters affect the visual control loop since the focal length describes the relation between image measurements and angles. If the lens zooms in part of the scene disappears from the image.

We can calculate the image Jacobian matrix, which describes changes of the distance between gripper and object in the image plane due to the changes of the focal length. (figure 6, b) Because of linear dependency of the distance on the focal length this Jacobian matrix does not depend on the focal length:

$$J_Z = \begin{bmatrix} \varphi_{33}(\alpha_L, \beta_L) & 0 \\ 0 & \varphi_{66}(\alpha_L, \beta_L) \end{bmatrix} \tag{7}$$

To realize the control algorithm described below we assume that pan and tilt angles are adjusted so that the midpoint between gripper and an object (figure 5, a) is in the centre of the image and the pan and tilt angles are known. The control loop is presented in figure 6, a.

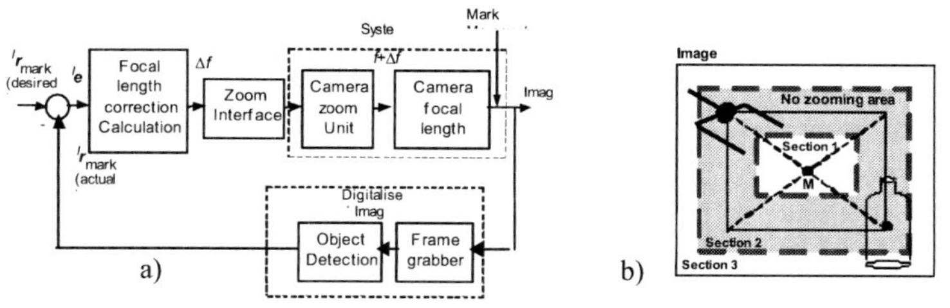

Figure 6: Zoom control loop and zoom areas

To prevent continuous zooming in and out, which strongly affects the control of the system, we denote the decomposition of the image in accordance to the necessity of zooming. The whole image is divided into three rectangular parts (figure 6, b). For each part we consider easy rules:

- if object and gripper are in section 1 (centre) → increase zoom;
- if object and gripper are in section 2 (grey) → no zooming
- if object and gripper are in section 3 (boundary) → decrease zoom

3 Obstacle avoidance

3.1 Trajectory generation

The movement of the robot arm can not be planned offline because in a service robot scenario the environment is not static, even if we assume that obstacles are not moving during a robot action. The system has to react online to different situations. If there are no obstacles between the gripper and the desired object no avoidance movement is necessary.

3.2 Common Trajectory generator for obstacle avoidance

The basis for collision avoidance in [3] is the epipolar constraint. The scene is observed with a stereo camera system and obstacles are detected with the help of image processing. With the knowledge of the epipolar constraint the two-dimension search of corresponding points is reduced to a simple one-dimension problem. The disadvantage of the epipolar constraint is the measuring of an exact calibration of the camera system.

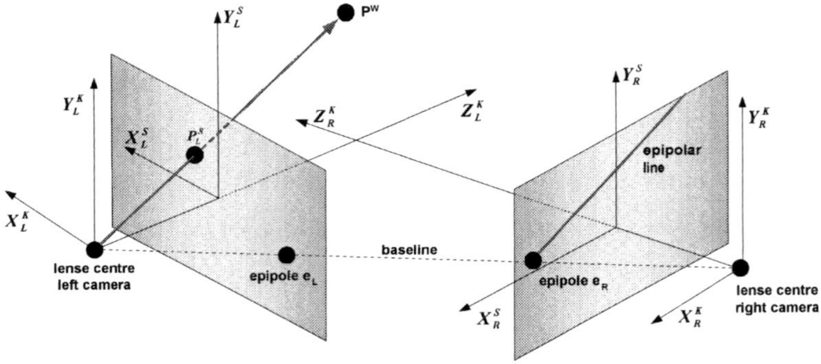

Figure 7: Epipolar constraint

Figure 7 shows a projection of a world point P^W from left to right camera. The point P_L^S in the left camera corresponds to a line in the image plane of the right camera. Such lines corresponding to points in the other image are called epipolar lines. All lines pass through the epipole e_R. The epipoles are points where the baseline intersects the image planes. The calculation of the epipolar constraints is given with the following equation

$$\left(P_L^S\right)^T \cdot E \cdot P_R^S = 0$$

$$\begin{bmatrix} u_L \\ v_L \\ 1 \end{bmatrix}^T \cdot \begin{bmatrix} e_{11} & e_{12} & e_{13} \\ e_{21} & e_{22} & e_{23} \\ e_{31} & e_{32} & e_{33} \end{bmatrix} \cdot \begin{bmatrix} u_R \\ v_R \\ 1 \end{bmatrix} = 0 \tag{8}$$

with P_L^S and P_R^S as homogeneous image coordinates. If the internal and external camera parameters do not change, the epipolar constraint E is constant. To determine the matrix E at least eight corresponding points are necessary. One method is to move the robot to eight different positions. From the detection of the arm tip we can set up eight independent equations. Another possibility is to confirm directly eight different points in one picture. Of course this calibration is accomplished offline only once. If now i.e. a point in the left image is measured $[u_L, v_L]$, the equation (3) has to be solved at $[u_R, v_R]$. The solution is an equation with two unknown variables, thus the point $[u_L, v_L]$ lies on a straight line $v_R = f(u_R)$ in the right image. With the help of this method the problem to find the corresponding point is highly simplified.

With the method of collision avoidance from [3] a straight line is inserted between actual und desired position of the end-effector in both images. If this straight line intersects an obstacle the line is divided into discrete steps (points in Figure 8). These points are transferred with the help of the epipolar constraint E to their epipolar lines in the right image. After that trajectory points of the right image are searched and will be moved along their epipolar lines until they lie outside of the detected obstacle. Figure 8 illustrates this method of construction of an avoidance trajectory in one image.

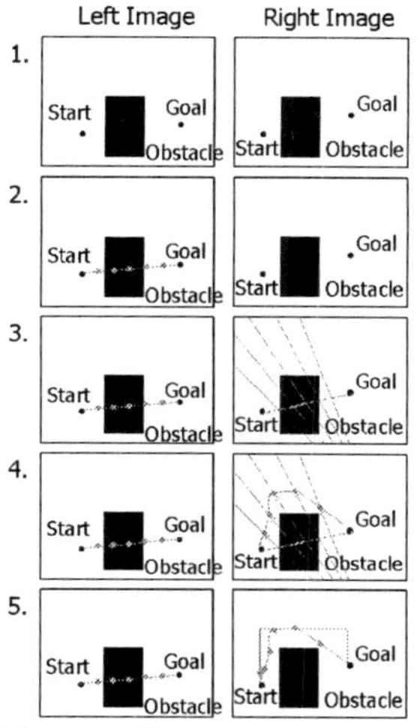

Figure 8: Trajectory generation

Step 1: Detection of start (gripper), goal (object) and obstacle

Step 2: Trajectory is a straight line between *Start* and *Goal* in the left image. The line is divided into discrete steps.

Step 3: With the help of the epipolar constraint, the epipolar lines corresponding to the points in the image 1 are projected onto the image 2.

Step 4: Move the points along the epipolar lines until they lie outside the obstacle

Step 5: To speed up the task, simplify trajectory to rectangular path.

3.3 Extended Trajectory generator for obstacle avoidance

With the method from 3.2 a fixed trajectory for obstacle avoidance can be calculated easily. But it can happen that the robot arm leaves the field of view i.e. through a huge avoidance movement. Another error could be a non-sufficient resolution of the cameras, i.e. the gripper could not be recognized in the image. That means the position of the cameras and the focal length has to be adapted. Thus the trajectory from Figure 8 (Step 5) is divided into discrete steps.

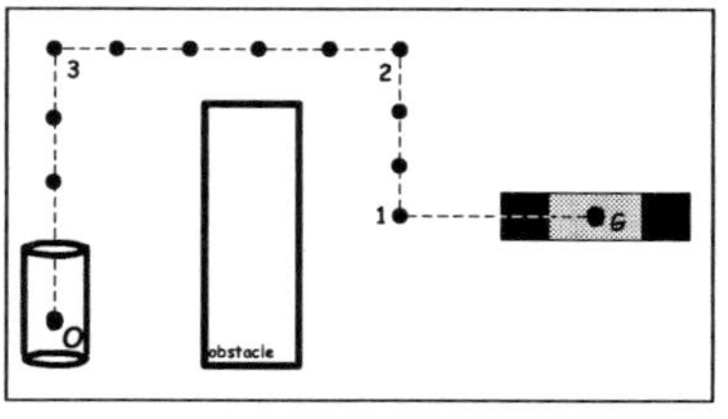

Figure 9: front view of the image

Figure 9 shows the provisional set points for the movement of the robot arm. A provisional set point is a virtual point without a detectable image feature in the real scene. The gripper and the object are represented with virtual points, i.e. the light emitting diode above the gripper has to be transformed to the centre of the gripper to determine the image based grasping point [2].

In Figure 9 we can see the gripper {G}, the object {O} and the obstacle. The gripper moves to point {1}. Then the obstacle is detected and an evasive motion leads to {2}. This procedure is repeated until the object {O} is reached. After every step the position of the gripper, the object and the obstacle in the image are checked to see whether they have a sufficing resolution. If a correction is necessary, calculated trajectory from 3.2 becomes incorrect, due to the changes in internal and external camera parameters. To speed up the calculation process the epipolar constraint matrix E is saved in advance for different focal lengths and different camera positions. Figure 10 shows the flow chart of this calculation.

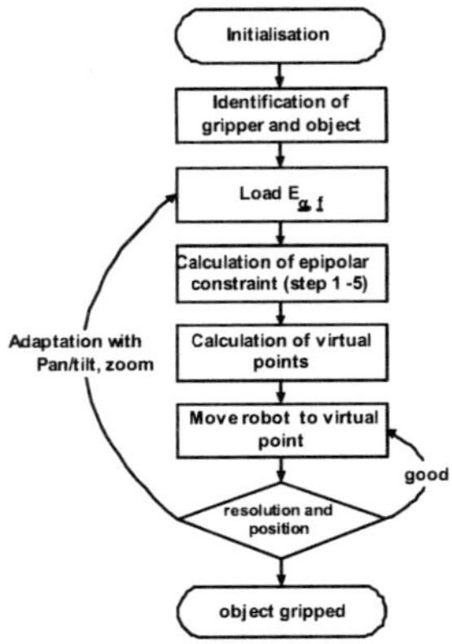

Figure 10: flow chart of calculation

4 Experiments

The typical task for the system FRIEND is to grasp a bottle or a glass from a tray.

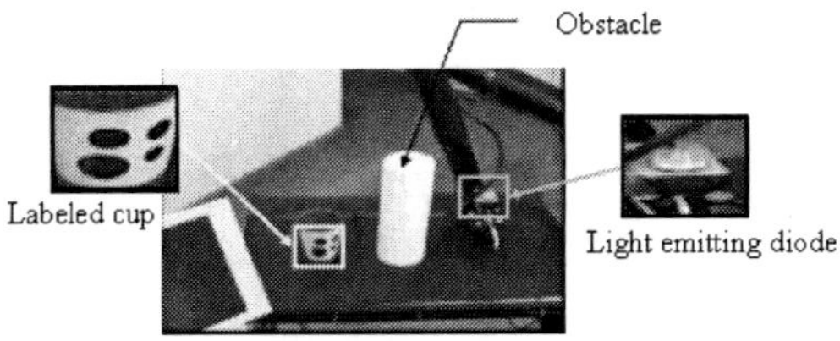

Figure 11: Detection of objects

Before starting the gripping process all objects of interest must be identified, i.e. the gripper and the object to be grasped. The other detected objects on the tray will be identified as obstacles. The glass to be gripped is labelled with four blobs. With the recognition of the object the set point for the control loop is known.

The position of end-effector is also necessary for each control step. The gripper is equipped with a blue light emitting diode (figure 11). The position of the end-effector is the actual control value.

Throughout all experiments the visual servoing task is to move the robot hand to a certain real-world point starting at an arbitrary point in the workspace. In the first experiment, the visual servoing with adjustable zoom cameras in a workspace free of obstacles is used. In the second experiment, the combination of the modified visual servoing method with the algorithm of obstacles avoidance is tested.

4.1 Visual servoing with adjustable zoom cameras

In Figure 12 the execution of an automatic gripping scenario is depicted. The task for the system is to grasp a glass in case an environment free of obstacles.

	Left Camera	Right Camera		Left Camera	Right Camera
1			2		
	$f=5.4$ mm; $\alpha=30^\circ$; $\beta=0^\circ$	$f=5.4$ mm; $\alpha=30^\circ$; $\beta=-5^\circ$		$f=12$ mm; $\alpha=30^\circ$; $\beta=0^\circ$	$f=10$ mm; $\alpha=30^\circ$; $\beta=-5^\circ$
3			4		
	$f=12$ mm; $\alpha=32^\circ$; $\beta=-2^\circ$	$f=10$ mm; $\alpha=35^\circ$; $\beta=-5^\circ$		$f=15$ mm; $\alpha=35^\circ$; $\beta=-5^\circ$	$f=15$ mm; $\alpha=37^\circ$; $\beta=-10^\circ$
5			6		
	$f=15$ mm; $\alpha=35^\circ$; $\beta=-5^\circ$	$f=15$ mm; $\alpha=35^\circ$; $\beta=-10^\circ$		$f=20$ mm; $\alpha=35^\circ$; $\beta=-5^\circ$	$f=20$ mm; $\alpha=35^\circ$; $\beta=-1\ 0^\circ$

Figure 12 Execution of grasping action with active zoom and pan-tilt control

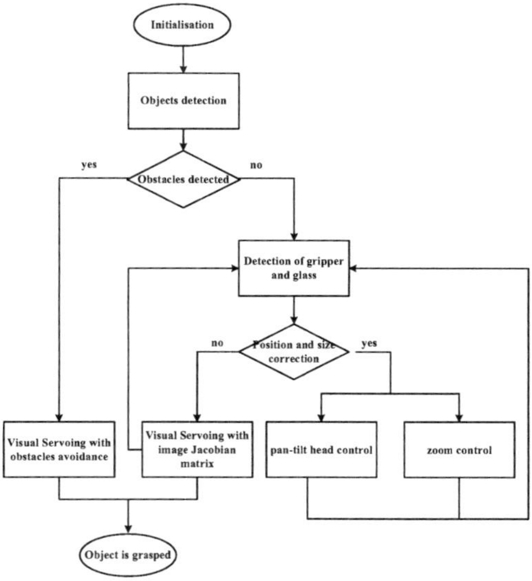

Figure 13 Flow chart

The start situation is presented in Figure 12, 1. The blobs on the glass are not detected because of their too small size and the focal length of both cameras must be changed. The focal length will be continuously increased until the blobs on the glass will be identified. As soon as both, the glass and the gripper, are found in both pictures the visual controlled gripping can be started (Figure 12, 2). According to the flow chart, presented in Figure 13, the system continuously tracks the gripper and the glass. The pan-tilt control holds the middle point between gripper and glass in the centre of the image and the zoom control adapts the focal length according to the rules, defined in section 2.3.2. If the gripper is in the grasp position (Figure 12, 4) the visual servoing algorithm will be switched off and the glass will be grasped automatically.

In the final control step the orientation and the focal length of cameras will be changed so that the glass is in the centre of the image and in good ‘size’ (figure 12, 6).

4.2 Obstacle avoidance with adjustable zoom cameras

Any objects between the gripper and the target are detected as obstacles. For the manipulator action, only obstacles in the neighbourhood of the manipulator are taken into consideration and the obstacle must be observable in both camera images. To simplify matters, it is assumed that the background is simple and the obstacle is distinguished sufficiently from the background by colour and that both can be described by rectangles. It is also assumed that those objects or at least parts of the obstacles are in the same plane as the gripper and the target.

To detect an obstacle a small rectangle area (region of interest, ROI) between gripper and object is defined. The area is bounded by the centre of gravity (COG) of the blobs at the target and the blue light emitting diode at the gripper (Figure 14). If any object is detected in this region in both images and the two ratios of object width to ROI width are equal, i.e. $w_{11} : w_{12} \cong w_{21} : w_{22}$, we can conclude that there is an obstacle between the gripper and the target object.

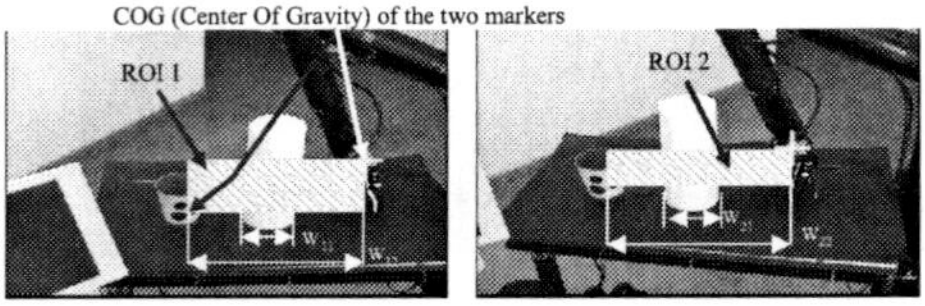

(a) Image1 (b) Image2

Figure 14. Defining the region of interest

Figure 15 shows the results of the experiment "gripping of a cup" with the robot arm Manus with use of visual servoing. Figure 15 (1) shows the calculated epipolar lines. The measured values for the rotation and the focal length of the cameras are:

$$\alpha_1 = \begin{pmatrix} \alpha_{L1} \\ \beta_{L1} \\ \alpha_{R1} \\ \beta_{R1} \end{pmatrix} = \begin{pmatrix} 35° \\ 5° \\ 35° \\ -5° \end{pmatrix} \quad \text{and} \quad f_1 = \begin{pmatrix} f_{L1} \\ f_{R1} \end{pmatrix} = \begin{pmatrix} 6{,}22 \\ 6{,}22 \end{pmatrix}$$

With this values the constraint matrix E is calculated:

$$\mathrm{E}_{\alpha_1, f_1} = \begin{bmatrix} 2.5\times10^{-6} & 3.1\times10^{-5} & -8.8\times10^{-3} \\ -2.6\times10^{-5} & 6.1\times10^{-6} & 1.2\times10^{-2} \\ 4.8\times10^{-3} & -1.5\times10^{-2} & 1.0 \end{bmatrix}$$

The robot arm moves along the trajectory until it reaches the position in Figure 15 (2). The position and focal length of the cameras are then adapted until the objects of interest are centred and have a sufficient resolution in the image Figure 15 (3). The trajectory from Figure 15 (1) does not longer hold and must be adapted with the new values

$$\alpha_2 = \begin{pmatrix} \alpha_{L2} \\ \beta_{L2} \\ \alpha_{R2} \\ \beta_{R2} \end{pmatrix} = \begin{pmatrix} 40° \\ 0° \\ 35° \\ -10° \end{pmatrix} \quad \text{and} \quad f_2 = \begin{pmatrix} f_{L2} \\ f_{R2} \end{pmatrix} = \begin{pmatrix} 7{,}37 \\ 8{,}04 \end{pmatrix}$$

for the new trajectory

$$E_{\alpha_2,f_2} = \begin{bmatrix} -9.1\times10^{-7} & -7.0\times10^{-5} & -4.8\times10^{-3} \\ 5.8\times10^{-5} & -2.2\times10^{-5} & 1.1\times10^{-2} \\ -1.0\times10^{-2} & -5.3\times10^{-3} & 1.0 \end{bmatrix}$$

This is repeated until the cup is reached, figure 15 (4).

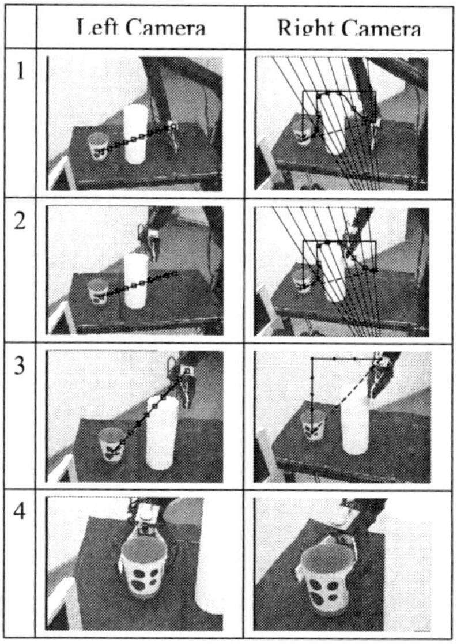

Figure 15: experiment

5 Conclusions

In this work we have shown the control method in an environment free of obstacles and in an environment with obstacles as well. The use of adjustable cameras in the control loop makes the system more flexible for different applications.

6 References

[1] G.D.Hager, W.C.Chang and A.S.Morse Robot Hand-Eye Coordination Based on Stereo Vision. *IEEE Control Systems Magazine vol.15, pp.30-39,1995.*

[2] P.I.Corke Visual Control of Robots: high performance visual servoing. Research studies press Ltd., 1996.

[3] T. Trittin and A. Gräser "The Method of Virtual Points for Autonomous, Image-Based Robot Control", *Proceedings of the 14th World Congress, IFAC*, pp. 293-8 vol.3, 1999

[4] K. Hosoda, K. Sakamoto and M. Asada "Trajectory Generation for Obstacle Avoidance of Uncalibrated Stereo Visual Servoing without 3D Reconstruction", *Proc. Of IEEE/RSJ Int. Conf on Intelligent Robots and Systems*, 1995.

[5] O. Faugeras Three-Dimensional Computer Vision, The MIT Press, 1993

First Experiences with Feedback Structures in Image Processing

Axel Gräser, Oleg Ivlev and Danijela Ristić
Institute of Automation, University of Bremen, 28359 Bremen
ag@iat.uni-bremen.de, oivlev@iat.uni-bremen.de and ristic@iat.uni-bremen.de

The basic idea of the inclusion of internal feedback structures in image processing is to make the results of image processing independent of external influences, like illumination conditions, characteristics of both the object and imaging system etc. In principle, feedback structures can be included in all phases of image processing in order to make the results of all processing units robust and reliable. The paper discusses main problems appearing by inclusion of feedbacks in digital image processing such as determining of feedback variables suitable from the control systems point of view. These variables have to be some measurable image characteristics determining the quality of images at each step of image processing.

The main emphasis in the paper is on the feedback between the quality of the original, not processed, object image captured by the camera and the object illumination condition. The better this condition is met, the original image will be of a better quality in the sense of its information content and, consequently the easier subsequent steps of image processing will be. Since a quality criterion is necessary to judge the illumination condition, the paper starts from the consideration of the image entropy as a measure of the average amount of information contained in the image. It is suggested to consider the values of image entropy in a dark, middle and light gray value area of the histogram as measures of the quality of original image. Then, the closed loop is built with the entropy of the image in a middle gray value area as the feedback variable. The controller uses the position of the light source as actuator variable. The paper also discusses the impact of this lowest closed loop on the edge detection, as one of the possible next steps of image processing needed for problems like character recognition, and, consequently on the generation of the OCR procedure inputs. The edge detection control loop is considered as well. Besides the consideration of inclusion of feedback principle in digital image processing, the paper shows some experimental results using the example of reliable character recognition tasks in car manufacturing.

1. Introduction

Digital image processing typically consists of two main parts: image acquisition and image data processing (Figure 1). Both of them are characterized by a low robustness [1,2].

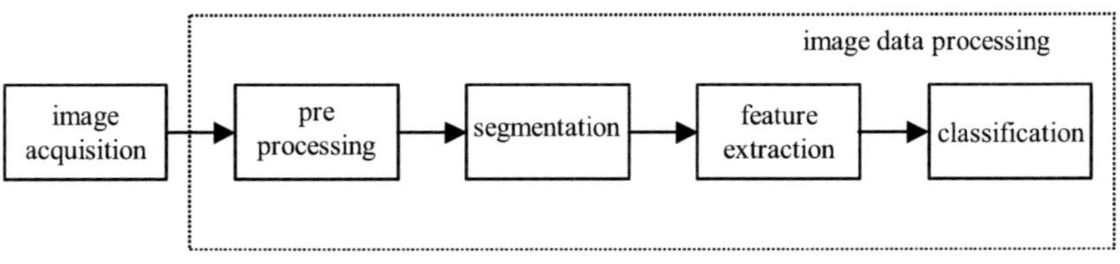

Figure 1. Block-diagram of open-loop digital image processing

In the phase of image acquisition the results are strongly influenced by illumination condition, object, lens system, characteristics of camera and frame grabber while in the phase of image data processing the results depend on parameters of algorithms chosen for preprocessing, segmentation, feature extraction and classification. Because of a high sensitivity of image processing to so many different influences, the adaptation of open-loop image processing to real-world visual inspection tasks is difficult and often gives unreliable results.

Consider the example of character recognition in car manufacturing where a reliable detection of identification numbers of the car body during the complete production process including the painting process is required. In practice there are different types of identification numbers. They can be stamped, founded, printed or just scratched on metallic surfaces (Figure 2).

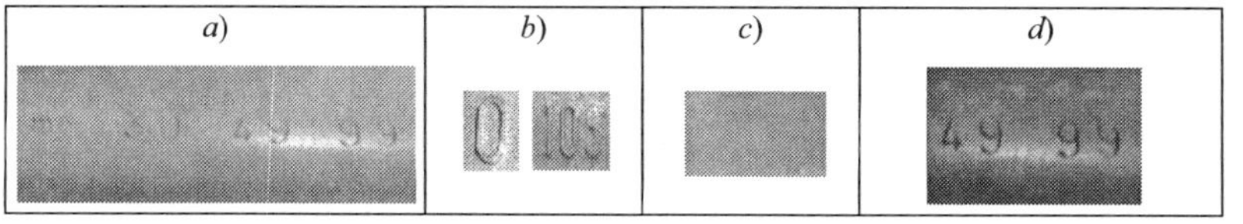

Figure 2. Different types of identification numbers:
a) printed *b*) founded *c*) scratched *d*) characters on curved metallic surface

As it can be seen in the case of founded numbers they can be damaged, in the case of scratched they are hardly visible. Besides this, the character detection is difficult because of additional disturbances as rust, oil etc. Different environmental conditions such as bad lighting and consequently undesirable reflection of light influence the process of character recognition to a great extend as well. Due to all these circumstances, the original images of the identification numbers are often of poor quality and thus not suitable for character recognition. But in spite of that, in many applications of image processing the engineers will regard the images as given and try to design algorithms that work on a stored set in order to change certain properties of the images, e.g. enhancing contrast, suppressing noise etc. This is time-consuming and often gives unreliable results. Therefore, we are of opinion that it is needed to think carefully about how the images are captured, in terms of lighting, lenses, camera position etc., and organise these factors before the acquisition in order to obtain good images. Also, it is needed to think about how the quality of the original images influences subsequent steps of image processing as well as how each previous step influences the next one. In order to achieve the main benefit of feedback in control systems, a high system robustness and disturbance rejection [8], we have considered the idea of feedback structures in all stages of digital image processing.

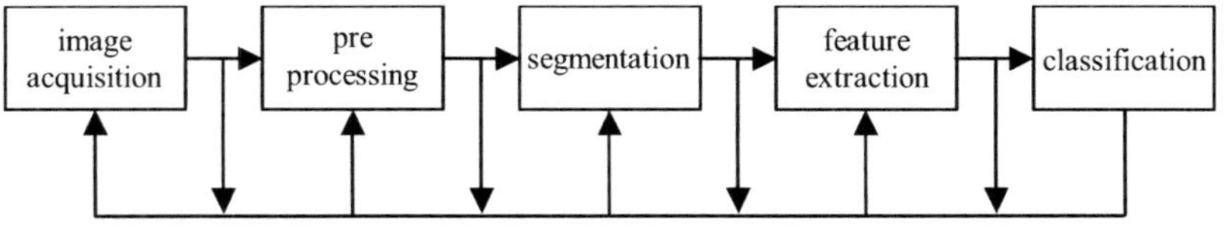

Figure 3. Block-diagram of closed-loop digital image processing

Besides the usual internal feedback structures aiming at the adjustment of the algorithm parameters [7], we have suggested feedbacks from the higher level of image processing to the lower level as shown in Figure 3. It is very important having in mind the fact that the steps of image processing are arranged sequentially. Hence, the poor and unreliable results of one processing unit may cause that the following processing step is almost impossible. In this hierarchical order, the image acquisition is the lowest level of image processing. Therefore, in our research we started from the image acquisition closed-loop.

2. Image acquisition closed-loop

It has already been mentioned in the introduction that for the reliable results of the recognition of identification numbers good quality of the original, not processed, object image is of major importance. The process of image acquisition is influenced by characteristics of camera and digitiser as well as the scene lighting conditions. Therefore, in several papers the control of the process the image acquisition has been investigated mostly from the adaptive camera regulation point of view. For example in [3] the sensor integration time, commonly called the "shutter speed", and digitiser characteristics gain and offset are considered as feed-forward control variables. In this paper we are focused on the application of image processing in the industrial processes. Since the most industrial processes are indoors it is practically possible to achieve controlled illumination conditions. Therefore we addressed ourselves to the consideration of illumination condition as decisive factor for the success of image acquisition. In this paper we concentrate on controlling the illumination condition and do not consider other aspects of acquisition such as camera and digitiser parameters or zoom and focus control. The feedback between the quality of original object image captured by camera and the object illumination condition is shown in Figure 4

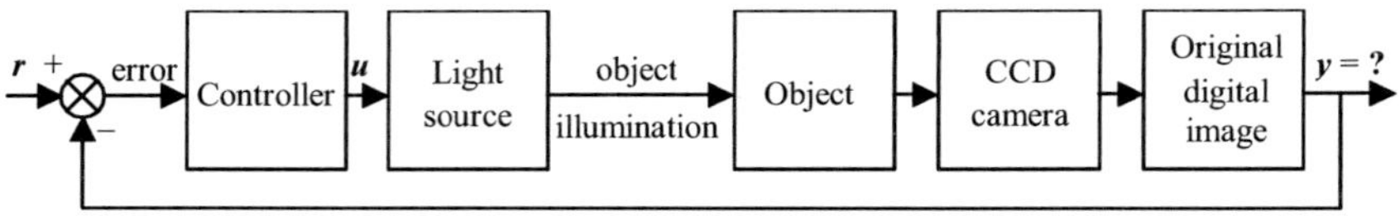

Figure 4. Feedback between the quality of the original object image and the object illumination condition

where ***y***, ***r*** and ***u*** represent respectively the controlled output, i.e. measurements determining the quality of the original image, the reference signal and the control signal being the parameters which determine the position of the light source and consequently illumination condition.

A basic requirement of the acquisition control is that the quality of the original object image has to be measured so that the control variables can be changed to optimise it. In our research some requirements have been set on the measure of image quality:

- it should be possible to calculate it easily from the image;
- it must be understandable what the image quality actually is;
- the measure representing the defined image quality must be found.

3. What is the image of good quality?

To find the answer to the question "what is the image of good quality?" is quite a difficult problem. The image quality obviously depends on the interpretation of the context. In the case of a "top model" image a good image hides some details like, for example, no perfect skin while for a surgical endoscope a good image is one showing the organ of interest in all details clearly for human interpretation. But the goal of our research has been to define a good image in the context of machine vision. It is known that practitioners in the machine vision area must keep in mind that "what human being can easily see is not at all simple for a machine" [5]. One of the main problems in the implementation of automated visual inspection systems is therefore to understand the way in which the machine "sees" and the conditions that have to be created for it to perform its task at its best. Since the subjects of our consideration have been the images of different types of characters on plates of different materials and colours, our research has been performed with one goal in mind: to achieve appropriate conditions of image acquisition resulting in an image suitable for image processing and leading to good machine perception and character recognition. The character recognition as one of the most complicated tasks in digital image processing consists of several processing steps having the same goal - to extract characters in symbolic form [6]. If the information content of the image has been lost during the image acquisition it will not be possible to recognize the characters by using any common image processing algorithm since the lost information can not be restored. In that sense, by authors opinion, the best measure of the original image quality in the context of machine vision can be determined under the condition that the subsequent image processing steps work well. And this will be the case if the original object image contains the maximum amount of information. Hence we have been motivated to look towards the information theory for the answer.

4. Entropy of an image as a measure of image quality

a) Entropy and information

In the classical information theory [4], the entropy of a information source which delivers L independent messages, each with a probability

$$p_i, \quad i = 0, \ldots, L-1 \tag{1}$$

is defined as the average information generated by the source and is given by:

$$H = -\sum_{i=0}^{L-1} p_i \log_2 p_i \quad \text{[bits/message]} \tag{2}$$

It is usual to use the base 2 logarithm in order to express information contained in the message in bits per message.

b) Entropy and image

Considering an image as a source with independent pixels, the entropy is defined as the information content of the image:

$$H = -\sum_{i=0}^{N-1} p_i \log_2 p_i \quad \text{[bits/pixel]} \tag{3}$$

where:

- p_i – probability of pixel value i

$$p_i = \frac{\text{number of pixels with gray level } i}{\text{total number of pixels in the image}}. \tag{4}$$

- N – number of pixel values (gray levels). For the usual case of an 8-bit integer image $N = 256$ when, according to (3) and (4), theoretical maximum entropy is 8[bits/pixel] corresponding to the case of all gray values equally distributed over the image pixels.

Having in mind the definition of image entropy (3) as the measure of the average amount of information, one can conclude that the maximum entropy means the image of best quality. The following discussion, based on the processing of test images of a metallic plate with characters on it, shows that it is correct but only to a certain degree. Images were captured in different conditions of illumination and then were processed. The image processing included at first the calculation of the original image entropy in order to see whether it behaved in a sensible way as the illumination condition changed. Then the binary edge-detected image was obtained in order to consider the relationship between the image entropy as possible measure of image quality and the results of edge-detection.

In the case of poor illumination condition during the image acquisition, the resulting image is almost totally dark (Figure 5(a)). Hence the corresponding image histogram, shown in Figure 5(b), is completely in the left part of histogram plane which means that almost all pixels have low gray values. Obviously it is impossible even for human beings to recognise any character in the image since its information content is very small and the entropy of the image is low: 4.77. Consequently the result of edge-detection is very weak and there is not one character in the binary edge-detected image (Figure 5(c)). If the object illumination is better, the visibility of the image will be enhanced as in the case of the image shown in Figure 5(d). So the information content of the image is a little bit larger and accordingly the image entropy is higher: 5.59. But, still this entropy is too small to have good results in the next image processing step and the binary edge-detected image (Figure 5(f)) contains only parts of hardly recognizable characters.

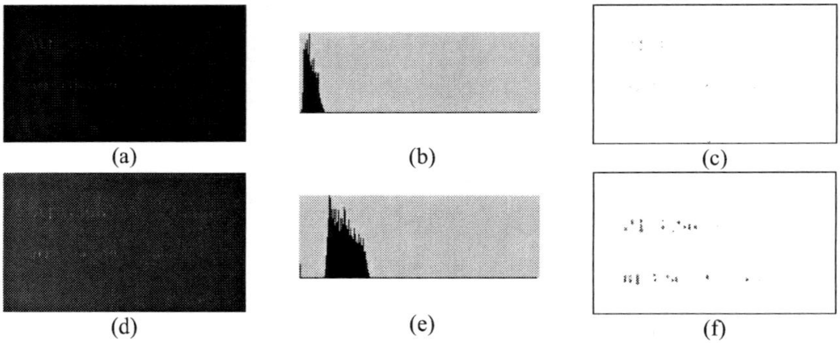

Figure 5. Left, original images showing the relationship between the entropy and the quality of an image. Middle, gray level histograms. Right, binary edge-detected images

If the illumination condition is better, the information content of the image and the entropy as its measure will be higher. But the largest entropy does not mean the best quality of the image in the sense of the best results of the next image processing step. The previous statement can be confirmed by Figure 6.

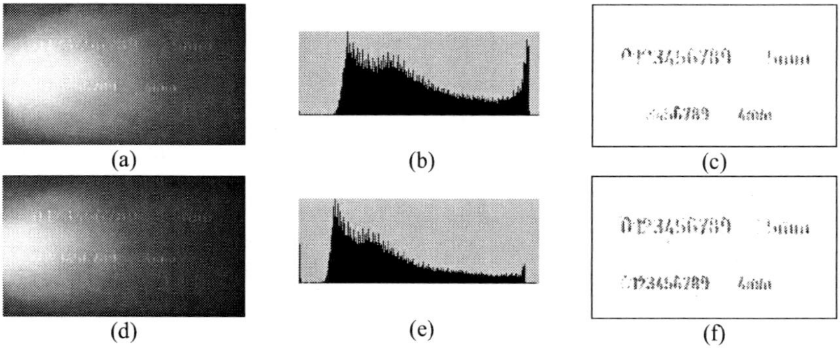

Figure 6. Left, original images showing that the largest entropy does not mean the image of best quality. Middle, gray level histograms. Right, binary edge-detected images

The original image shown in Figure 6(a) has the largest entropy achieved in the performed test: 7.58. As it can be seen from the image histogram that has a peak in the right bright part of the histogram plane, the large number of pixels have very high gray value (equal or near 255). This indicates the existence of a bright area in the original image covering some characters which so can not be detected in the process of edge-detected image binarization (Figure 6(c)). Evidently, the binary edge-detected image which corresponds to the original image shown in the Figure 6(d) with lower entropy in comparison with the previous one: 7.51, has larger number of detected characters. It is to be expected since the image histogram (Figure 6(e)) does not have a peak in the light gray value area of its scale which means that the bright part does not hide the characters to be recognized. On the basis of previous remarks it turns out that the maximum entropy could not be a good measure of image quality. We can only say that the image entropy should be as large as possible or it should be maximal but under the some constraints.

In order to determine the mentioned constraints we have to examine first the undesired cases. There are two. The first one is the case of a totally dark image as in the case of the image shown in Figure 5(a) and the second one is the case of image that is too bright as the one shown in Figure 7(a). Obviously the considered image consists of a bright area spreading across the image. This causes the lowering of entropy to 5.36 which, from own side, indicates a loss of information. Hence, as it can be seen from the corresponding binary edge-detected image in Figure 7(c), no character at all can be detected. What can be said is that the two undesired cases are images with a large number of pixels that have either values that are too small or too high. It turns out that the desired case is the image with the biggest number of pixels having the gray value from the middle area of the gray scale. This is the case of the image shown in Figure 7(d) which has 94.42(%) of total number of pixels in the middle area of the gray scale. Hence, the distribution of number of pixels in different areas of the gray level scale is very important for the quality of the image.

Starting from the results of our research we have been suggested the division of the gray scale into three areas: dark, determined by a gray levels interval 0 – 35; middle, determined by a gray levels interval 36 – 179 and light corresponding to a gray levels interval 180 – 255.

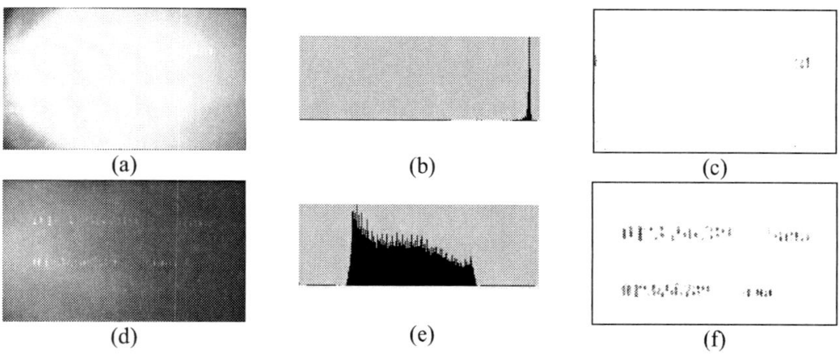

Figure 7. Left, original images showing the relationship between the number of pixels having different gray values and the information content of the image. Middle, gray level histograms. Right, binary edge-detected images

On the basis of previous remarks one can ask "why do we need to speak about entropy when we can simply speak about the distribution of number of pixels?". But the example in Figure 7(d) shows that only the biggest number of pixels in the middle gray value area can not be a good measure of image quality. Namely, the high number of pixels in the middle gray value area causes a histogram that is too narrow. The corresponding image is of low contrast which reduces the entropy to 7.13 and indicates a loss of information. As a result the characters in binary edge-detected image in Figure 7(f) are hardly recognizable. Therefore, the measure of image quality can not be expressed as "to have as large number of pixels in middle gray value area as possible", it has to be connected with the stretchiness of the image histogram. Entropy is such measure due to its connection with the probability of occurrence of pixel intensities. It results from the definition of the image entropy (3) that if the probability of observing the gray value is small, a large degree of information can be concluded, since the occurrence of this gray value is rare. Contrary to that, gray values with a large probability of occurrence, which is the case of narrow histogram, have a very small degree of information. Hence, for the purpose of good image quality it is of major importance to have the image histogram stretched over the whole gray scale but with the largest amount of information carried by pixels from the middle gray value area which is the case of image shown in Figure 8(a).

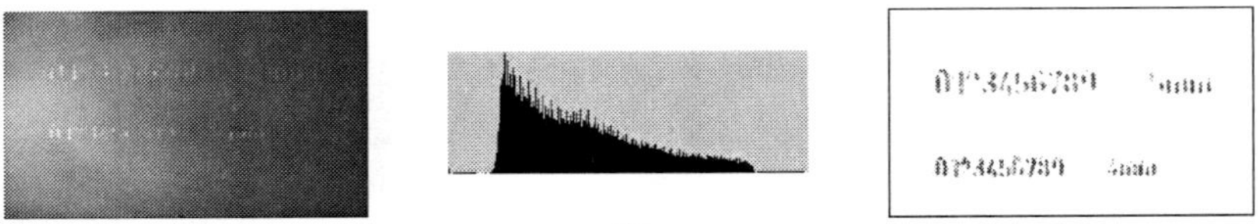

(a) (b) (c)

Figure 8. Left, original image with desirable values of entropy in different gray value areas. Middle, gray level histogram. Right, binary edge-detected image

As a result of our research we have suggested the desired distribution of image entropy in three gray value areas as shown in Figure 9.

Gray value areas:	**Dark** (0 .. 35)	**Middle** (36 .. 179)	**Light** (180 .. 255)
Desired values of entropy:	0.15 .. 0.35	6.4 .. 6.7	0.4 .. 0.95

Figure 9. Desired value of entropy in different gray value areas

All images with the value of entropy in dark, middle and gray value areas from the concluded desired intervals will be of a good quality. The quality will be provided in the sense that even in the case of uneven illumination (objects are illuminated only from one side as in the case of our experiment) the image will not have parts that are either too dark or too bright making the correct character recognition impossible and the image histogram will be stretched over the full gray scale. As it has been already mentioned, such an image is shown in Figure 8(a). The corresponding histogram indicates that the image has a good contrast which provides a good base for character recognition.

Since test results show that near-optimal performance is obtained with the desired values of image entropy in dark, middle and light gray value areas of the histogram, we have concluded that these measures are suitable for control of image acquisition when the definition of image quality is that the subsequent image processing step should work well. This conclusion can be confirmed by Figure 10 showing the plots of entropy in different gray value areas of original images vs. the number of black pixels in the binary edge-detected image.

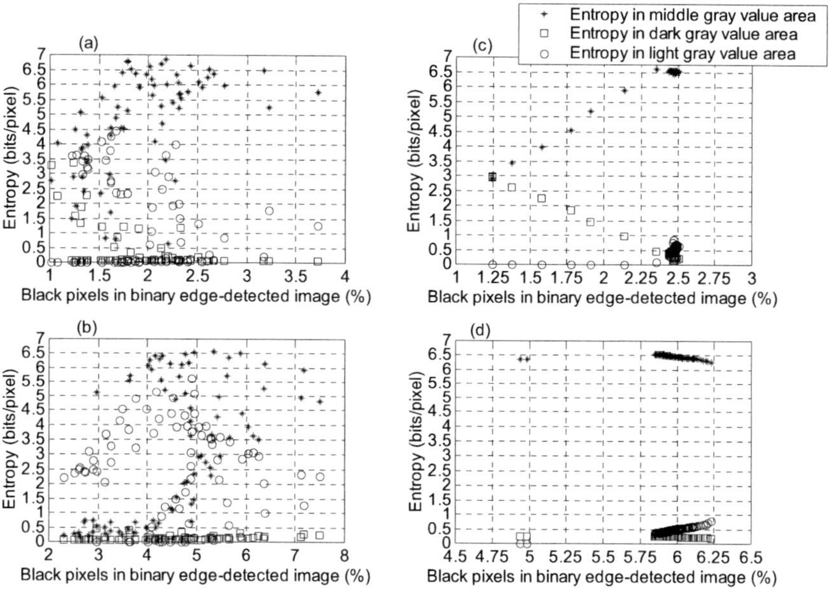

Figure 10. Plots of entropy in different gray value areas of original images vs. the number of black pixels in binary edge-detected images. Wide (left) and narrow (right) entropy range

In order to validate our conclusion that the values of entropy in different gray value areas can be used as a measure of the original image quality under the condition that the subsequent image processing steps work well, we should have looked for a measure of quality of the binary edge-detected image. Having in mind the nature of the problem of character recognition we have considered the number of black pixels in a binary edge-detected image as the proper measure of quality. At first sight it seems that for the good results of character detection the biggest number of black pixels is needed. However, the results shown in graphs 10(a) and 10(b) indicate that the biggest number of black pixels corresponds to original, not processed, images with a large entropy in light gray value areas. Hence some of these pixels are "not useful" since they correspond to "false" edges resulting from improper illumination condition. But, it is apparent that if the image has the desired combination of entropy values in dark, middle and light gray value areas then the number of black pixels will be quite large and, which is more important, these pixels belonging to the characters to be detected will be useful from the character recognition point of view. This is more clear with the graphs 10(c) and 10(d) showing the relationship of entropy in different gray value areas and the number of black pixels in binary edge-detected images for a narrow entropy range. Results shown in Figure 10 demonstrate that near-best performance, which can be expressed as "binary edge-detected image should have as big number of black pixels as possible", is obtained when the original image has entropy values in three gray value areas belonging to desired intervals. Therefore it has been concluded that these measures are appropriate for image acquisition control in the sense that the obtained image can be used as right input to the edge detection image processing algorithm as one of the next steps in the character recognition process.

5. Experimental results

The inclusion of the feedback principle in digital image processing and the handling with the image entropy value in different gray value areas as feedback variable for image acquisition closed-loop has been investigated through an experiment performed at the IAT laboratory of service robotics. The objects – metallic plates of different colours with characters on them - have been imaged in conditions of different illumination. The changeable illumination conditions have been provided by changing the light source position held by the robot arm. The light source position is determined by three parameters: distance between the light source and the object, and sweep and tilt of the light source with respect to the object (Figure11).

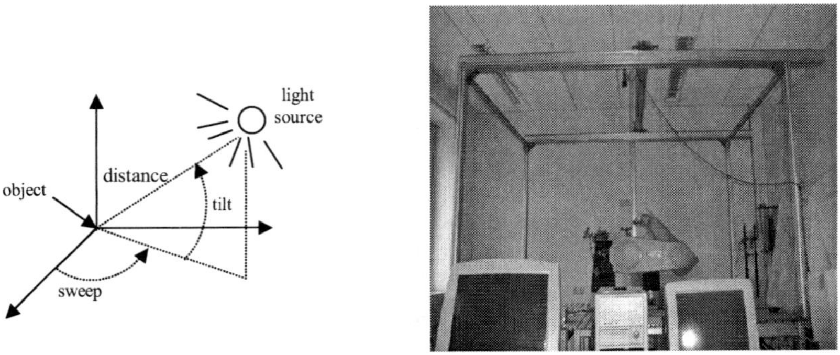

Figure 11. Light source position with respect to the object.
Left, principle scheme. Right, the system set in IAT service robotics laboratory

Throughout our research we considered sweep and tilt as constant and the distance between the light source and the imaged object as the only changeable parameter determining the light source position. Therefore the output of the implemented P-control algorithm has been a distance signal determining the appropriate position of the light source and consequently the optimal illumination condition, and the input has been an error signal of the entropy value in the middle gray value area. The block–diagram of the image acquisition closed-loop with implemented P-control law is shown in Figure 12

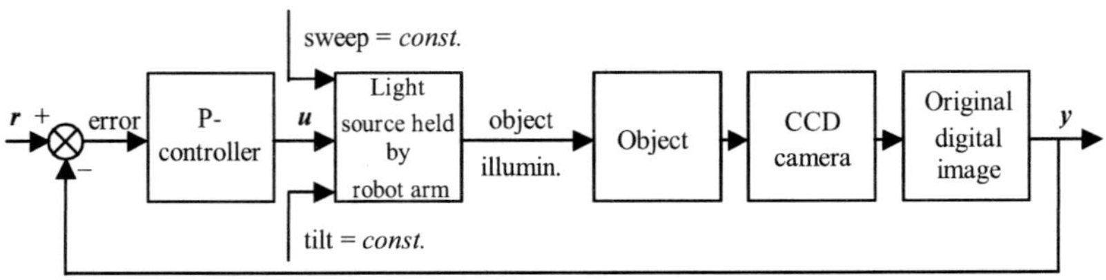

Figure 12. Image acquisition closed-loop with implemented P-control law

where ***y***, ***r*** and ***u*** represent respectively the controlled output (entropy in the middle gray value area as measure of the original image quality), the reference signal (desirable value of entropy in middle gray value area) and the control signal explained above.

The control results achieved with the implemented P-controller is surprisingly good, considering its simplicity, the fact that the P-gain was determined by "trial and error" method and the fact that plant dynamics is pure dead-time due to the image processing. Figure 13 shows the closed-loop response in the case of a gray plate as imaged object.

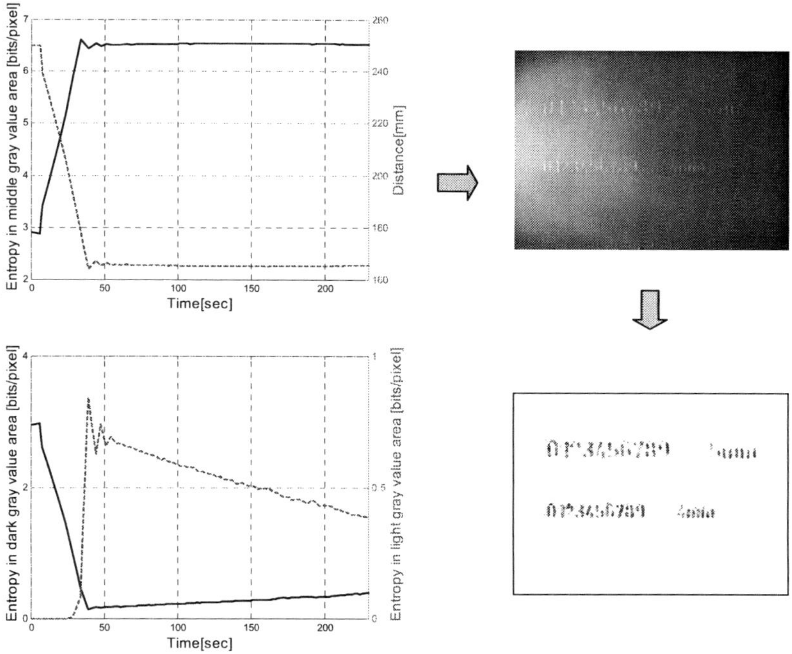

Figure 13. Left, closed-loop response: left upper, entropy in middle gray value area (solid line) and distance (dash line); left bottom, entropy in dark gray value area (solid line) and entropy in light gray value area (dash line). Resulted original image (upper right) and corresponding binary edge-detected image (bottom right)

A specific characteristics of implemented feedback structure for image processing purposes distinguishes it from the usual industrial control. Here it is not needed to keep the controlled output constant in time. For the purpose of further image processing leading to the reliable character recognition (OCR algorithm) results only one “good” image should be chosen as it is shown in Figure 13. It is needed to emphasize that even if only the entropy value in the middle gray value area was controlled by the implemented SISO controller, the entropy values in dark and light gray value areas were investigated so that the chosen resulting image has the desired combination of entropy values in all three gray value areas. Hence, this image is “good" input for the edge detection closed-loop. The block–diagram of the edge detection closed-loop is shown in Figure 14 where:

- y_2 is the ***controlled output*** – number of black pixels in binary edge-detected image
- r_2 is the ***reference signal*** – desired number of black pixels in the binary edge-detected image.

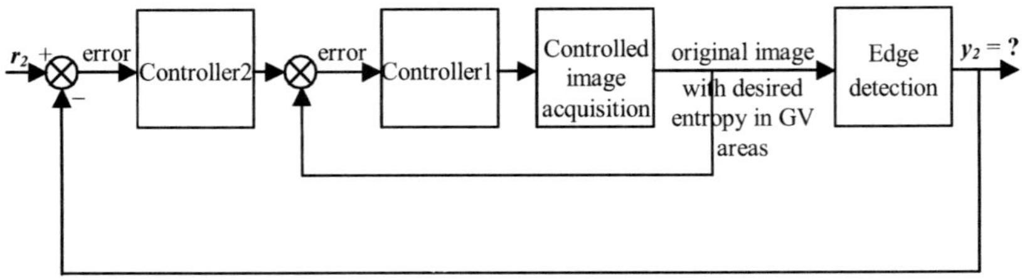

Figure 14. Edge detection closed-loop

The feedback variable of the edge detection closed-loop has been chosen on the basis of our research results shown by Figure 10. Besides the appropriate number of black pixels in the resulting binary edge-detected image it is of bigger importance that these pixels are coherent giving the "full" characters. It is obvious from Figure 13 that the detected characters in the binary edge-detected image are not "full" due to the object illumination from one side only determined by a constant sweep value. In order to reach "full" characters to be detected and so the "ideal" input for the OCR algorithm a possible next step in our research could be the merging of images resulted from the edge detection closed-loop as it is shown in Figure 15. In further step, the control-loop using the number of recognized characters as controlled variable and the sweep angle as actuator signal, could be built.

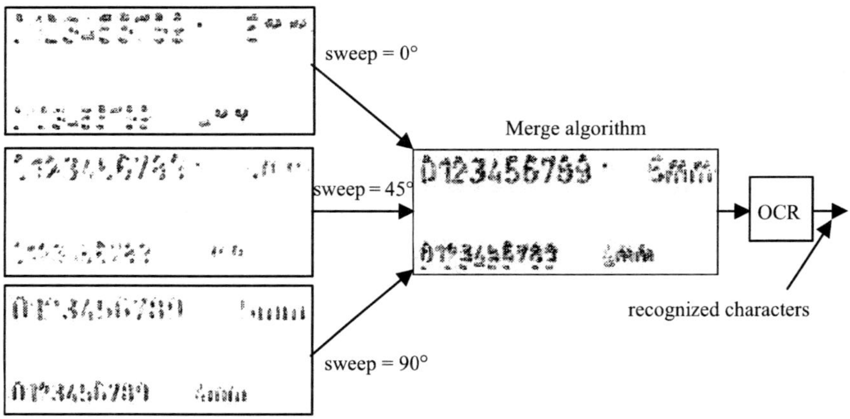

Figure 15. Possible next step: Merging of images in order to make "ideal" input for OCR

6. Summary and Conclusions

- The research results presented in this paper have confirmed that entropy values in different gray value areas of the image histogram can be used as a measure of the original, not processed, image quality for image acquisition control where the definition of the image quality is that subsequent image processing steps should work well. However, the considered entropy as a variable only relating to the image histogram does not take into account the spatial properties of the image, which is an aspect of particular importance in image analysis. Therefore in future research we will consider other image quality measures providing better insight in spatial image characteristics which can be used as feedback variables.

- It has been shown that with the implemented simple P-control algorithm the image acquisition control results are quite good. Apart from that the future research work will be directed to the improvement of the image acquisition control loop and consequently to the improvement of the edge-detection and the character recognition control loop.
- One of the main parts of our future work will be further experimentation, research and improvement of the image processing algorithms needed for the character recognition task.

7. Literature

[1] Gonzales R. C., Woods R. E., Digital Image Processing, Prentice-Hall, 2002.

[2] Jähne B., Practical Handbook on Image Processing for Scientific Applications, CRC Press LLC, 1997.

[3] Marchant J. A., Testing a measure of image quality for acquisition control, Image and Vision Computing 20, pp. 449-458, 2002.

[4] Maitre H, Zinn-Justin J., Entropy, Information and Image, Progress in Picture Processing, , Springer-Verlag, pp. 81-115, 1996.

[5] Demant C., Streicher-Abel B., Waszkewitz P., Industrial Image Processing-Visual Quality Control in Manufacturing, Springer-Verlag Berlin Heidelberg, 1999.

[6] Mori S., Nishida H., Yamada H., Optical Character Recognition, John Wiley & Sons, 1999.

[7] Gräser A., Martens C., Rehabilitation Robots Transfer of Development and Research Results to Disabled Users, Proceedings of 3rd International Workshop on Human-friendly Welfare Robotic Systems HWRS 2002, pp. 99-107, 2002.

[8] Aström K. J., Hägglund T., PID Controllers – Theory, Design and Tuning, Instrument Society of America, 1995.

Image processing function for compensation of low-contrast images acquired during welding process

Dorin Aiteanu, Axel Gräser
Institute of Automation, University of Bremen,
Otto-Hahn-Allee NW1, D 28359 Bremen, Germany
Emails: aiteanu@iat.uni-bremen.de, ag@iat.uni-bremen.de

Abstract

To improve the quality of the images recorded in the welding environment with a high-dynamic-range video camera, an automatic brightness/contrast adjustment procedure is combined with a non-linear transformation having the characteristics of the Bezier curve. This permits quality improvement for images before and during welding, which have different illumination characteristics and gradients. A maximal enhancement of contrast is reached and the image in the brightest area is preserved, performing a smooth transition. The welding arc image is preserved and the work scene is better represented.

1. Introduction

The purpose of the TEREBES[1] project is to design an intelligent welding mask for the observation of the welding arc, in order to improve the manual welding conditions. The system has to provide a better view of the work environment, additional welding parameters during the welding process and to elaborate documentation for certifying the weld seam quality.

The images are acquired with two HDRC (high-dynamic-range-CMOS) cameras [1] for three-dimensional stereoscopic representation. They are processed using a wearable computer for improvement of the quality, to provide a good view of the welding scene. Afterwards, they are merged with additional data, like the actual welding machine parameters (voltage, current), and also data extracted from the acquired images (position and inclination of the welding torch, identification of the work piece, etc.) The processed images and additional data are displayed to the user on a stereoscopic Head-Mounted-Display (HMD), integrated with the cameras in a new developed welding helmet [2].

2. Image Quality

To develop an algorithm which improves the images before and during welding, the properties of the HDRC camera are used. The complex characteristics of the images has been resumed in two simple parameters, i.e. the contrast and brightness coefficients. To detect the presence of the welding arc, a filter has been developed, which switches between the linear and nonlinear transformation functions.

2.1 High Dynamic Range Camera

To acquire images having a strong illumination gradient (very bright areas with very dark areas) and still preserve the details in both domains a HDRC-camera is used which performs a logarithmic compression of the light intensity. Due to this fact, the acquired images are of 'low-contrast' (Fig.1) [2].

If the images are processed with the same parameters of a linear contrast stretching and brightness adjustment filter, neither the bright nor the dark area are preserved. To enhance the image in the dark area will lead to saturation in the dark area and vice versa.

---

[1] TEREBES is a research project sponsored by the German Ministry for Education and Research, www.terebes.uni-bremen.de.

a) b)

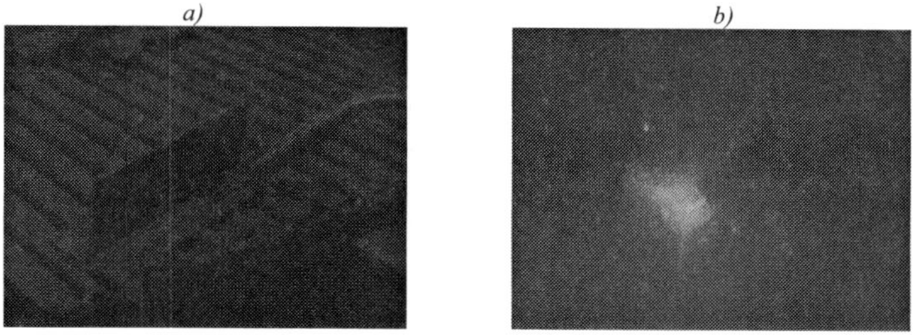

Fig. 1 The input image, provided by the HDRC camera:
a) before welding; b) during welding

Fig. 2 shows the processing of images before and during welding using the same parameters for brightness and contrast in both cases. While the improvement of the image acquired before welding was successful, the information regarding the welding arc is eliminated by saturation.

2.2 Automatic Brightness / Contrast Adjustment

a) b)

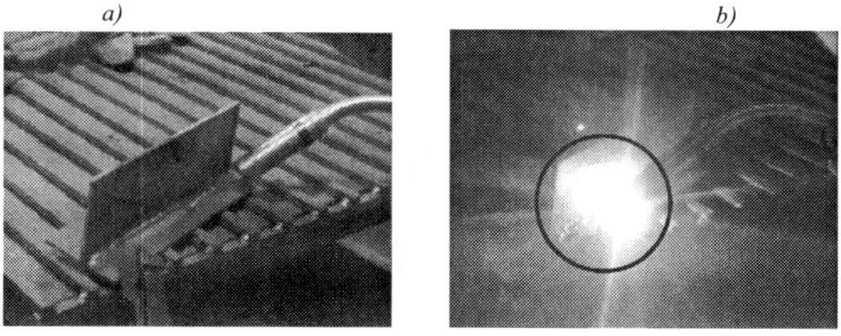

Fig. 2 Processing of images before and during welding
with the same brightness / contrast parameters
a) before welding; b) during welding

The brightness of the image represents the amount of light contained, characterising brighter and darker images. On the other side, the contrast shows the way how the range of the possible intensity vales has been used to distinguish the details in the image. To achieve a mathematical description of the above mentioned parameters, much easier to be determined in the automatic processing, the following considerations has been made.

2.2.1. The Brightness

The **brightness** of an image could be described using the median value of the pixel intensities p_{med}, (with other words, the intensity value of one pixel in the image, having the same number of brighter pixels and darker pixels) divided by the range of the intensity values.

Considering P_{min} and P_{max} the minimal and maximal value of the greyscale, respectively (i.e. for 8 bit greyscale images $P_{min} = 0$ for black and $P_{max} = 255$ for white), the brightness characterising an image is described using the following formula:

$$b = \frac{p_{med}}{P_{\max} - P_{\min}} \tag{1}$$

The brightness coefficient is a real value in the interval *b0 [0, 1]* describing a darker image for a bright image the coefficient is close to 1.

2.2.2. The Contrast

The image **contrast** can be described as the difference between the quartile 0.75 and 0.25 of the intensity values, divided by the range of the intensity values.

$$c = \frac{q_{0.75} - q_{0.25}}{P_{\max} - P_{\min}} \tag{2}$$

The contrast coefficient, a real value in the interval $c\ 0\ [0,\ 1]$ describes a rich contrast image for values close to 1 and a low contrast image if the coefficient is close to 0. For example, a binary (threshold) image will have the contrast coefficient c = 1.

2.2.3. Optimizing the Image

To determine the correction parameters in order to obtain an optimal output image, valid for various characteristics of the output image, an algorithm has been developed, based on the following considerations.

- The ***brightness coefficient*** is defined as optimal, id b is close to 0.5.

$$b_{def} = 0.5 \tag{3}$$

If the acquired image has values far from b_{opt} a linear point operation is defined [5] with:

$$p_{o(x,y)} = p_{i(x,y)} + B_{cor} \tag{4}$$

where the B_{cor} is the absolute brightness correction, defined as:

$$B_{cor} = b_{cor} \cdot (P_{max} - P_{min}) \tag{5}$$

and the relative brightness correction is calculated as:

$$b_{cor} = (b_{def} - b) \tag{6}$$

Variables $p_{o(x,y)}$ and $p_{i(x,y)}$ are the pixel values in the output and input images respectively (Fig.3).

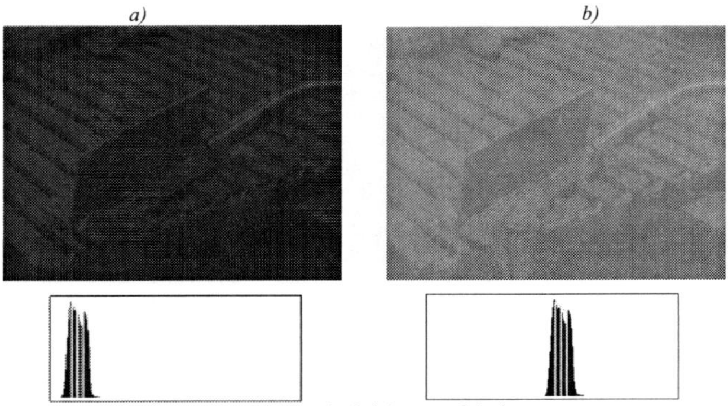

Fig. 3 Adjusting the brightness of the image
a) input image and histogram ; b) output image and histogram

Particularity

For the special case like the welding process, the most important visual information is concentrated in the brighter area of the image. By using the above mentioned for computing the brightness correction, it could occur that the welding arc information will be saturated (Fig. 4).

It can be easily observed that the "distance" between the median of the input image and the proposed position (middle of the greyscale) is greater than the difference between the information of the welding area and the maximal admitted grey value. In this manner, the pixel values which are virtual over the maximal admitted value (see Fig. 4, a) are limited by the maximal value, producing the saturation effect (Fig. 4, b).

To detect the domain of the grey values of the welding arc, the first solution was to simply detect the maximum value of the histogram, considering that during welding, the arc is the brightest element in the picture. Practical, this method is very sensitive to noise, and will neutralize the effect of the brightness correction.

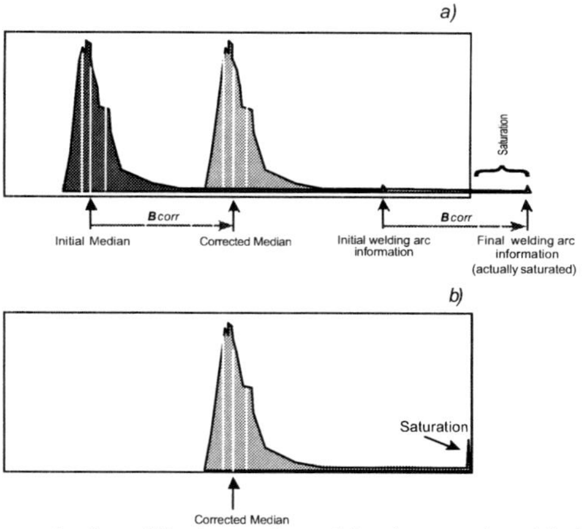

Fig. 4 Effect of saturation for welding images, by applying the correction of the brightness; *a) the information of welding arc is eliminated by shifting over the maximal grey value; b) the histogram of the processed image.*

In order to make the algorithm stable and not sensitive to image noise, an algorithm has been developed using the saturation index s. This is described by the amount of saturated pixels ($N_{sp} \mid p_{(x,y)} = P_{max}$, pixels having the maximal possible value) divided by the total number of pixels in the image N_{tp}.

$$s = \frac{N_{sp}}{N_{tp}} \tag{7}$$

The virtual maximum value, which depends on the saturation index M_{sat} has been defined as:

$$M_{sat} = q_{(1-s)} \tag{8}$$

where $q_{(1-s)}$ is the quantile *(1-s)* from the input image pixel values, and represents actually the lowest pixel value in the input image which will be saturated in the processed image.

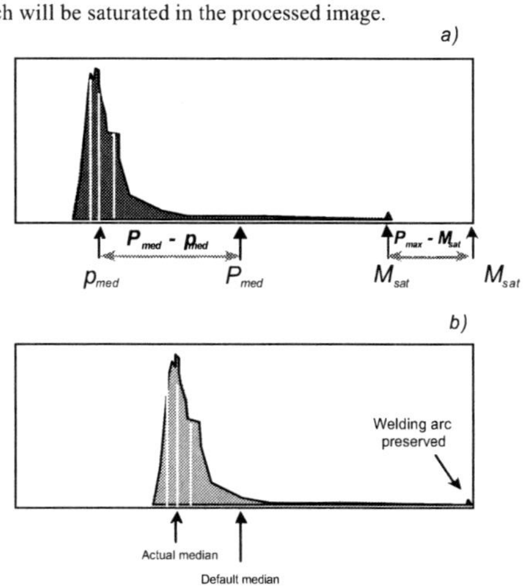

Fig. 5 Adjusting the brightness with preserving the welding arc image; *a) emphasise the differences; b) histogram of the processed image*

Considering this (see Fig. 5), the enhanced coefficient for brightness correction is described as:

$$B_{cor} = \min\left((b_{def} - b)(P_{\max} - P_{\min}),\ (P_{\max} - M_{sat}) \right) \tag{9}$$

With the final transformation (5), the brightness of the image is optimized preserving the welding arc and meanwhile setting the median histogram value as close as possible to the default median.

- The ***contrast coefficient*** should have the "ideal" value c = 1 to provide the maximum contrast. This requirement is not applicable for natural images, because applying the maximum possible contrast enhancement, will lead inevitably to a lack of information due to saturation. Considering the special requirements in the welding process observation, where the important information is concentrated in the brighter area of the image (welding arc), the contrast has to be stretched with the maximum value while avoiding saturation.

The contrast correction C_{cor} is determined using the contrast stretching formula [5]:

$$p_{o(x,y)} = (p_{i(x,y)} - a_0)\frac{b_1 - a_1}{b_0 - a_0} + a_1 \tag{10}$$

where b_0 and a_0 are the maximum and the minimum pixel value of the input image, respective b_1 and a_1 of the output image. Considering that the pixels with the light intensity equal to M_{sat} (already containing the brightness correction B_{corr}) should have the maximum intensity value P_{max} after transformation, the initial maximum and minimum (b_0 and a_0) are considered equal to P_{max}, respective P_{min} and the final maximum and minimum (b_0 and a_0) are obtained by adding the contrast correction offset C_{cor} to the initial maximum P_{max} and by subtracting from the minimum P_{min}.

$$\begin{aligned} p_{o(x,y)} &= P_{max} \\ pi(x,y) &= M_{sat} + B_{cor} \\ a_0 &= P_{min} \\ b_0 &= P_{max} \\ a_1 &= a_0 - C_{cor} = P_{min} - C_{cor} \\ b_1 &= b_0 + C_{cor} = P_{max} + C_{cor} \end{aligned} \tag{11}$$

By replacing (11) in (10) results:

$$P_{max} = (M_{sat} + B_{cor} - P_{min})\frac{P_{max} + C_{corr} - (P_{min} - C_{cor})}{P_{max} - P_{min}} + P_{min} - C_{cor} \tag{12}$$

To simplify the calculations, P_{min} is considered 0 (the minimum normal value for digital image data). After the calculations (from 12), the contrast correction C_{cor} results:

$$C_{cor} = \frac{P_{\max} \cdot (P_{\max} - (M_{sat} + B_{corr}))}{2 \cdot (M_{sat} + B_{corr}) - P_{\max}}, \tag{13}$$

Remark: Good results have been obtained with a saturation index $s = 0.1 \cdot 10^{-3}$ (Fig. 6).

a) b)

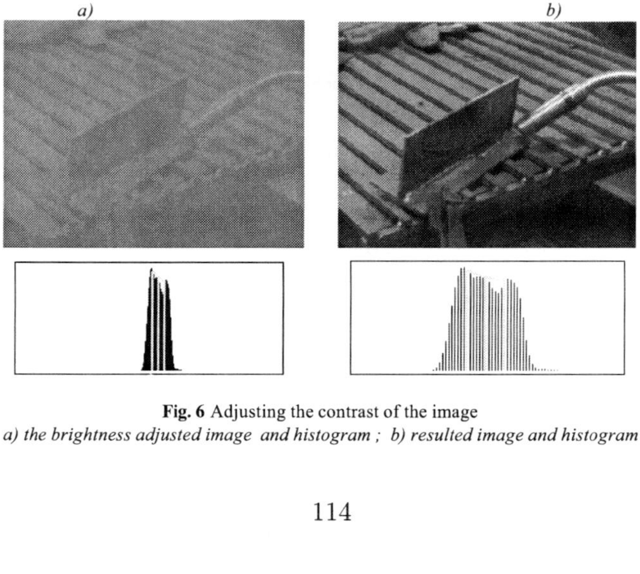

Fig. 6 Adjusting the contrast of the image

a) the brightness adjusted image and histogram ; b) resulted image and histogram

2.2.4 Results

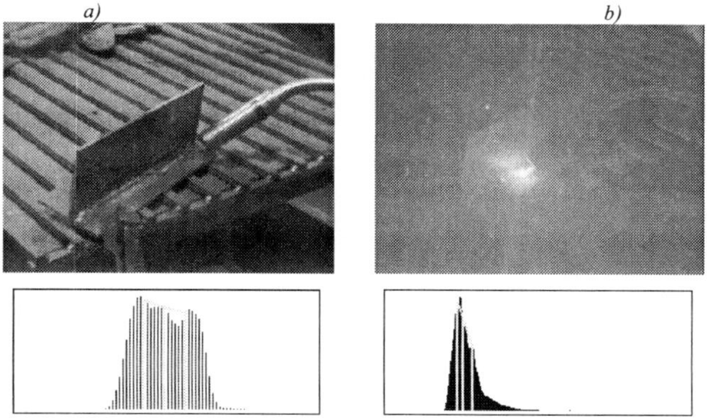

Fig. 7 Applying the algorithm: the image before welding is improved (a), while the image during welding is preserved (b)

The transformation for the brightness and the contrast adjustment is described (using (4) and (10)) as:

$$p_{o(x,y)} = (p_{i(x,y)} + B_{cor})(1 + \frac{2 . C_{cor}}{P_{max}}) - C_{cor} \qquad (14)$$

By the use of this algorithm, the brightness and the contrast of the input images are adjusted, offering good results for the images before the welding process, but preserving the welding arc in the case of images acquired during the welding process (Fig. 7).

2.3 The Non-linear Transformation

The purpose of the automatic brightness/contrast adjustment presented above is to offer a better image *before* the welding process, in order to extract important data from the image for augmentation (e.g. the position and inclination of the welding torch, identification of the welding piece) and/or to offer a better view to the user. By using the algorithm, the images acquired *during* the welding process are mainly preserved unchanged. In order to obtain an adaptive contrast improvement in the background of the image during welding (to extract data to augment the visual information), but still preserving the welding arc, a non-linear transformation has to be developed.

2.3.1. The Bezier Curve

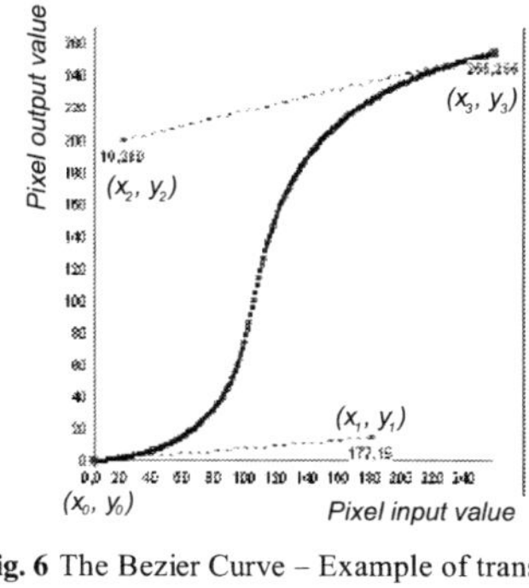

Fig. 6 The Bezier Curve – Example of transformation

Bezier curves (Fig.6) represent simple cubic equations[2]. A cubic Bezier curve is defined by four points: (x_0,y_0) is the *origin* point and (x_3,y_3) is the *destination* point. The points (x_1,y_1) and (x_2,y_2) are *control points*. Two equations define the curve: $x(t)$ and $y(t)$, with t = 0..1. The equations[3] are:

$$x(t) = a_x t^3 + b_x t^2 + c_x t + x_0, \qquad (11)$$
$$y(t) = a_y t^3 + b_y t^2 + c_y t + y_0,$$

where the coefficients a_x, b_x, c_x and a_y, b_y, c_y are determined based on the four points of the curve:

$$\begin{aligned} c_x &= 3\,(x_1 - x_0) & c_y &= 3\,(y_1 - y_0) \\ b_x &= 3\,(x_2 - x_1) - c_x & b_y &= 3\,(y_2 - y_1) - c_y \\ a_x &= x_3 - x_0 - c_x - b_x & a_y &= y_3 - y_0 - c_y - b_y \end{aligned} \qquad (12)$$

To generate the values for x and y, t takes values from 0 to 1 with a certain step. The smaller the step the more points from the curve will be calculated and the curve is more precise.

Fig.6. presents the principle of the transformation. The horizontal axis represents the pixel values of the input images, while the vertical axis describes the values of the pixels in the output images, conform to the curve.

2.3.2. Principle of the Bezier Transformation [4]

The saturation effect must be avoided, which results normally from standard linear transformation like contrast stretching. This is the main purpose of using the nonlinear Bezier transformation. This was, the pixels in the input image with values between the minimum and maximum value (P_{min}, P_{max}), will remain in the same domain. To cover the whole domain of the light intensity, the minimum and maximum value of intensity must not exceed the limits. This means that the start and the end point will be (P_{min}, P_{min}) and (P_{max}, P_{max}) respectively. The shape of the Bezier curve will be influenced only by the position of the control points.

Fig. 7 presents the principle of the Bezier transformation, and the effect on the image histogram. If the transformation will be made with respect to the normal line (the first bisector), the output pixel will have the same intensity value as the input pixel and the image will remain unchanged.

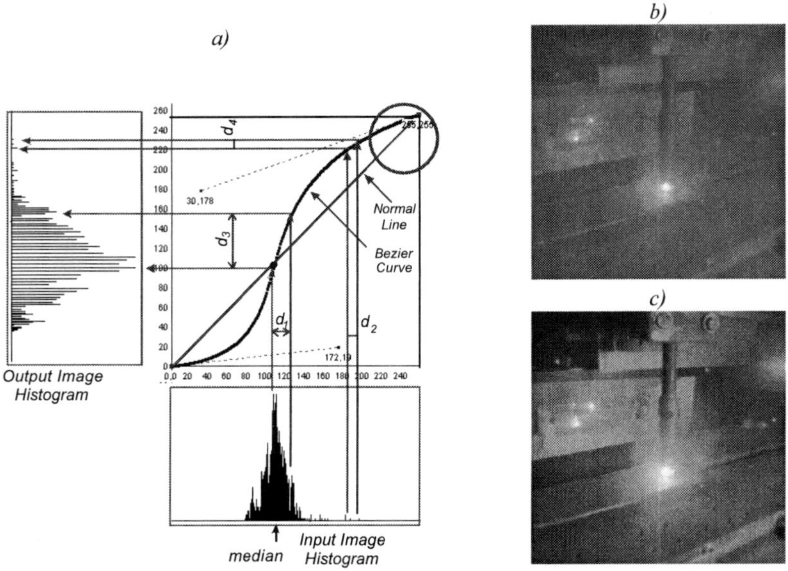

Fig. 7 The Bezier Transformation Principle:
a) the Bezier Curve and histograms; b) the input image; c) the output image

[2] Bezier Curves - developed by Pierre Bezier in 1970 for CAD/CAM operations.
[3] Equations conform to Adobe Post Script Reference, Adobe Corporation, 2000

With the nonlinear Bezier transformation curve marked in figure 7, the image histogram will be stretched more around the median value and less by approaching to the domain limits. For a better comparison, the effect on two sections d_1, d_2 in the input image is that the difference in intensity for two points around the median value is considerably bigger in the output image (see Fig. 7.a, mark d_3) than in the input image (d_1). For some points in the proximity of the maximum value the distance is nearly the same, possibly lower (compare distances d_2 with d_4).

Remarks:

- although the background is better represented, the image of the welding arc is still influenced (Fig. 7a) marked with circle)
- the coordinates for the control points (x_1,y_1) and (x_2,y_2) are difficult to be determined automatically, based on the input image characteristics.

2.3.3. Transformation using double Bezier curve

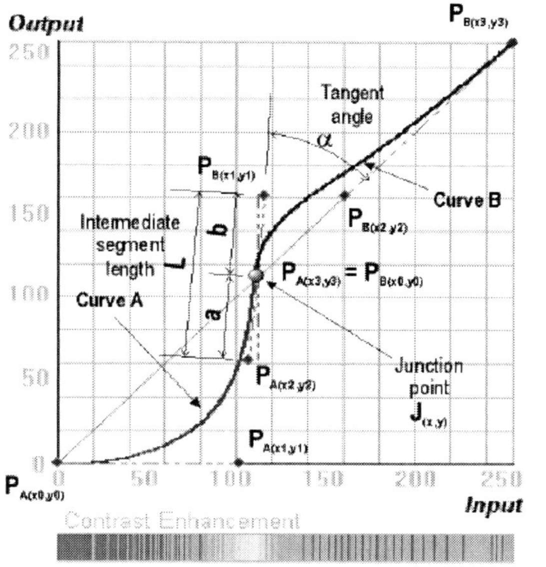

Fig. 8 Special Bezier transformation with contrast enhancement bar.

To accomplish the above mentioned requirements for selective contrast improvement in the image, a nonlinear function consisting in two jointed Bezier curves (Fig. 8) is defined.

The curve is composed by two different Bezier curves (A and B) being connected by the junction point $J_{(x,y)}$ (with $x = y$), which influences the magnitude of contrast stretching in the image. The angle a is defined by the angle between the line from P_{min}, P_{min} (0,0) and P_{max}, P_{max} (255,255) and the tangent to both curves in the tangent point. For a = 45°, the contrast stretching is maximal around the junction point. In figure 8 the effect is displayed with a contrast enhancement bar. Light grey values in the bar show that the contrast is locally enhanced, dark grey areas show that the contrast is reduced. For fine adjustment, another two characteristics are introduced, namely the length of the segment (L) which determines the position of the intermediate points as well as the distribution of this length to each curve, defined as ratio R = a/L.

While the values for the ratio R and the length L are experimentally determined and than fixed, the values for the angle and junction point are calculated using some characteristics of the input image.

It has been observed, that the useful information from the surroundings of the welding arc is concentrated around the quartile $q_{(0.25)}$ of the histogram. Experiments have demonstrated that the contrast improvement around this value leads to the best results. To evaluate how much the contrast has to be locally stretched, the measure $\Delta q = p_{med} - q_{(0.25)}$ is used, in the following equation:

$$A = A_{\max\_ref} \cdot \left(1 - \frac{\Delta q}{\Delta q_{ref}}\right) \qquad (13)$$

where A_{max_ref} represents the maximal allowed value for the tangent angle (namely $A_{max\_ref} = 45°$), while the constant Δq_{ref} is the desired distance for Δq after transformation, in light intensity units. This means, that if Δq is lower than Δq_{ref} in the image, a double Bezier curve will be generated which will transform the image in order to achieve a distance for Δq closer to the reference.

2.3.4. Computing the control points for the curve

The start, end and control points are determined for the two Bezier curves, as described in *Table 1*.

Table 1 – The determination of the control points for curves A and B

Curve	***Point***	***Observation***
A	$P_{Ax0} = P_{min} = 0$; $P_{Ay0} = P_{min} = 0$;	*The curve A starts at the point coordinates (P_{min}, P_{min}), confirming that the information in the lower intensity area will not be reduced to the minimal value.*
	$P_{Ax1} = J - (1 - \tan(45-a))$; $P_{Ay1} = P_{min} = 0$;	*To produce a smooth transition of lower intensity values to the minimum value, the control point $P_{Ax1,\,Ay1}$, (related to the start point $P_{Ax0,\,Ay0}$) is located at the intersection of the horizontal axe and tangent line (T)*
	$P_{Ax2} = J - L \cdot R \cdot \sin(45-a)$; $P_{Ay2} = J - L \cdot R \cdot \cos(45-a)$;	*The control point $P_{Ax2,\,Ay2}$ related to the joint points is located on the tangent line (T) at positions determined by the segment (L) and ratio (R).*
	$P_{Ax3} = J_x = q_{(0.25)}$; $P_{Ay3} = J_y = q_{(0.25)}$;	*The end point of curve A has the same coordinates as the start point of curve B and points to the intensity value in the picture where the contrast stretching will be maximum. (In this case, quartile q0.25 of the histogram)*
B	$P_{Bx0} = P_{Ax3} = J_x = q_{(0.25)}$; $P_{By0} = P_{Ay3} = J_y = q_{(0.25)}$;	
	$P_{Bx1} = J + L \cdot (1-R) \cdot \cos(45+a)$ $P_{By1} = J + L \cdot (1-R) \cdot \sin(45+a)$	*The control point $P_{Bx1,\,By1}$ related to the joint points is located on the tangent line (T) at positions determined by the segment (L) and ratio (R).*
	$P_{Bx2} = P_{By1}$; $P_{By2} = P_{By1}$;	*To generate a smooth adjustment of the Bezier curve , the control point relative to the end point belongs to the first bisector (P_{By1} , P_{By1})*
	$P_{Bx3} = P_{max}$ $P_{By3} = P_{max}$	*The curve B ends at the point (P_{max}, P_{max}) ensuring that the information in the higher domain of the histogram will NOT be saturated*

The influence of the parameters *J*, *a*, *L* and *R* on the shape of the curves, was emphasised in figure 9. It can be observed by following the contrast enhancement bars under each graph, the domain of the greyscale where the contrast is improved.

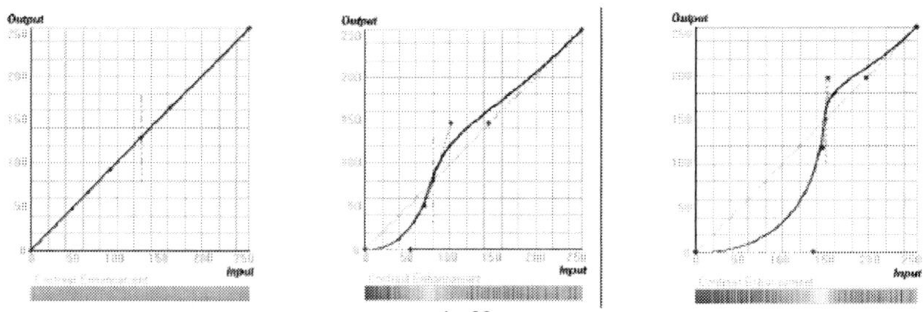

J = 127, a = **0 °**, L = 100, R = 0.5 | J = 80, a = 27 °, L = 100, R = 0.3 | J = 150, a = 40 °, L = 80, R = 0.4

Fig. 9 Influence of the parameter onto the shape of the curve

This represents the main advantage of the Bezier transformation, where certain domains of the intensity values can be improved by means of contrast, while other areas are maintained unchanged. For the special case like observation of welding process, the welding arc area of the image is preserved, while the range of the pixels in the area of low intensity is emended, offering the possibility to extract useful data, like object edges, visual markers etc.

3. The Processing Structure

The images before welding, which don't contain a strong light source are improved with good results with the linear transformation, described in chapter 2.2. The images acquired during welding contain strong illumination and are successfully processed with the nonlinear procedure described in 2.3 to enhance the details in the surroundings of the arc. An algorithm has been developed to detect the probable state of the scene and to switch to the appropriate improvement function. The advantage is to save the computing time needed for the nonlinear transformation of the images before welding. Here only the linear correction fulfils the need of improvement.

3.1. Switching between filters

The algorithm for the automatic image improvement in the two specific situations - before and during welding uses a property of the HDRC camera, namely logarithmic light compression. In this way, the difference between the maximum (based on saturation index) and the median of the image histogram D has values on a certain domain for images before the welding process, where no direct light source is present (Fig. 10).

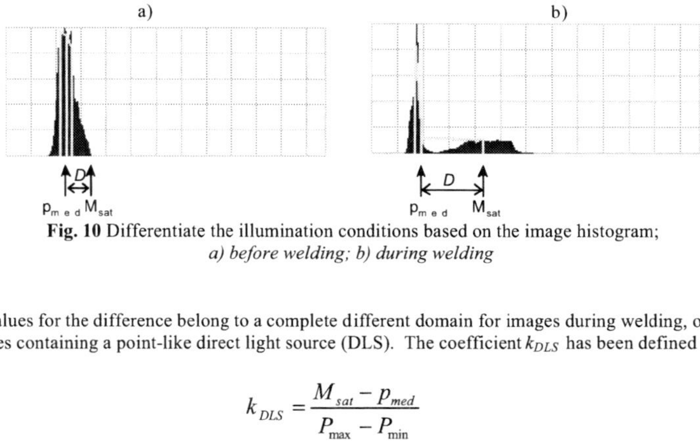

Fig. 10 Differentiate the illumination conditions based on the image histogram;
a) before welding; b) during welding

The values for the difference belong to a complete different domain for images during welding, or in general – for images containing a point-like direct light source (DLS). The coefficient k_{DLS} has been defined as:

$$k_{DLS} = \frac{M_{sat} - p_{med}}{P_{max} - P_{min}} \tag{14}$$

which has values in the interval of k_{DLS} 0 (0.10, 0.25) for images without a direct light source, and k_{DLS} 0 (0.35, 0.70) when a welding arc source is contained. Considering this, the reference value for k_{DLSref} = 0.30 has been chosen to switch to the nonlinear transformation.

3.2. The camera control-loop

The camera electronic allows the setting of two parameters - gain and offset to adjust the characteristic of the camera. By adjusting the offset, the mapping to 8 bit representation of image information under various illumination conditions could be adapted. The following control loop has been developed which adjusts the camera offset based on the median value of the acquired image.

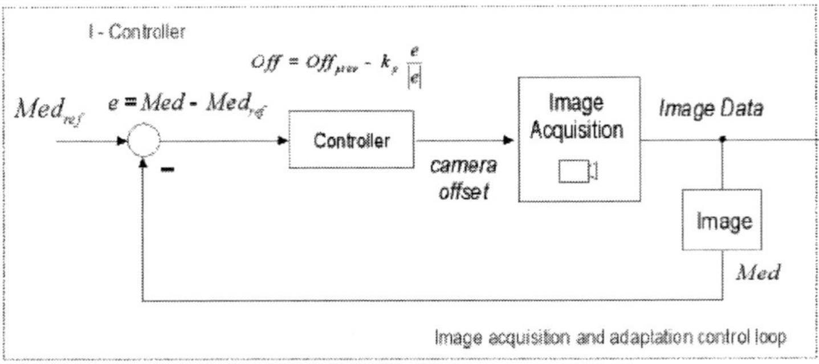

Fig. 11 Camera control loop

The image is acquired using a default offset. Afterwards, the median pixel value *Med* is extracted from the image, and compared with the reference Med_{ref}. The difference is used to calculate the offset correction for the next image $?k = k_p \frac{e}{|e|}$, which is subtracted from the previous camera offset Off_{prev} in order to determine the new offset *Off*. In order to achieve a stable offset adjustment, which should not be influenced by the fast changing of light conditions especially during the impulse welding, the value selected for the coefficient $k_p = 1$.

3.3. The Algorithm

The algorithm for image improvement contains the camera offset control loop, followed by the automatic linear contrast/brightness improvement based on the requirement of preserving the welding arc. Afterwards, the non-linear Bezier transformation is applied, in the case when a direct light source is detected (Fig.12).

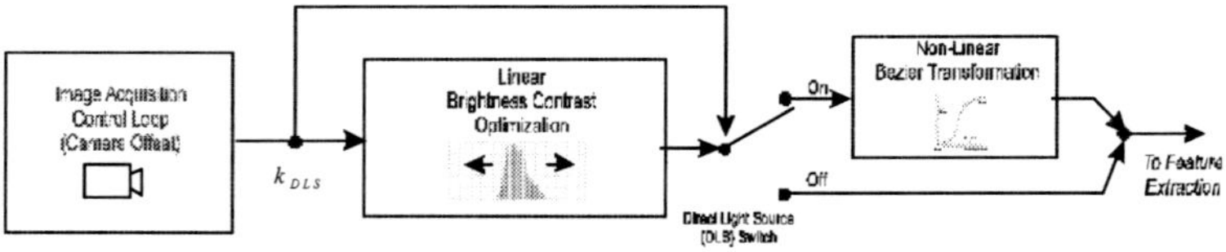

Fig.12. The structure of the algorithm

To emphasize the effects of the algorithm, a simple simulation stand has been prepared containing a light source and a plate with markers of various sizes and positions, which is illuminated only by the faint ambient light and not by the light source. Under two different illumination conditions, (A - without a direct light source and B - with light source) intermediate images in the processing sequence has been captured. Table 2 shows the acquired image (1), the image after linear brightness-contrast improvement (2) and the final image, after the nonlinear Bezier transformation (3). Direct light source coefficient k_{DLS}, the brightness *b* and contrast *c* parameters are displayed for each case.

When no direct source is detected, only the linear transformation is performed (Table 2, section A-2), producing a severe increase of the brightness (from 0.19 to 0.49) and contrast (from 0.04 to 0.23). The nonlinear transformation is in this case bypassed. When the light source is present (detected using the k_{DLS}) the linear transformation produces an improvement of the brightness, while the nonlinear transformation enhances the contrast in the surroundings of the light source, while maintaining the light source image area unchanged (compare histogram for B-3 with B-2).

Table 2 – Comparison of the effects of the algorithm to the image information

	Image Acquiring (1)	Linear Brightness-Contrast Optimization (2)	Non-linear Bezier Transformation (3)
(A) No Light Source	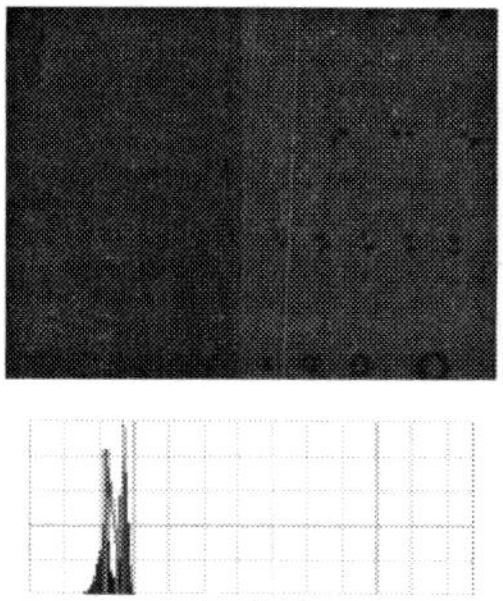 $b = 0.19$ $c = 0.04$ $k_{DLS} = 0.126 < k_{DLSref}$	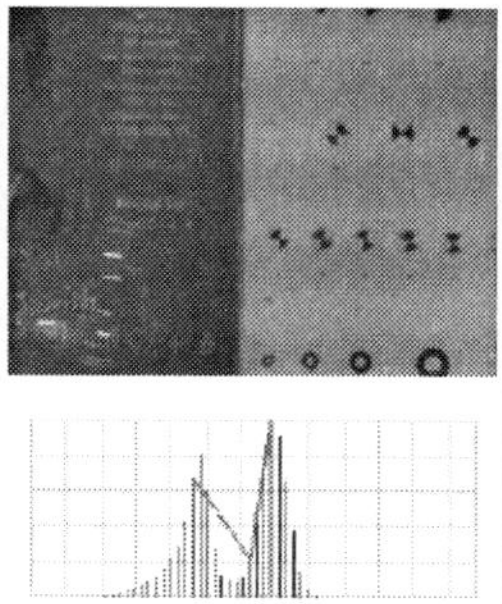 $b = 0.49$ $c = 0.23$	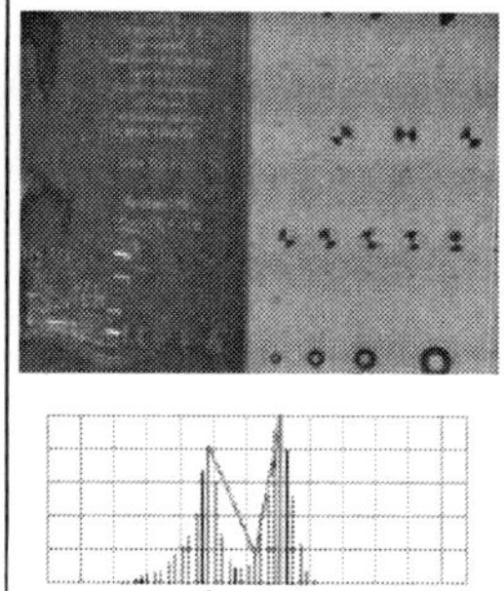 $b = 0.49$ $c = 0.23$
(B) Light Source	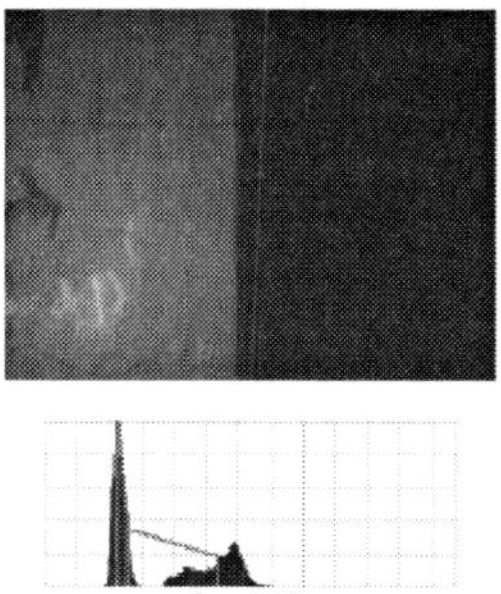 $b = 0.19$ $c = 0.24$ $k_{DLS} = 0.616 > k_{DLSref}$	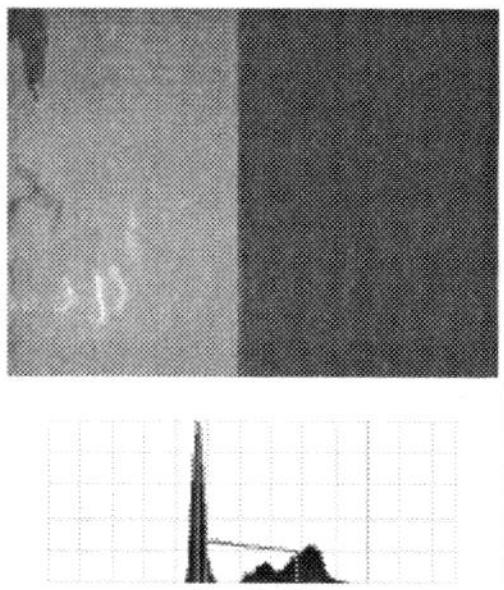 $b = 0.38$ $c = 0.24$	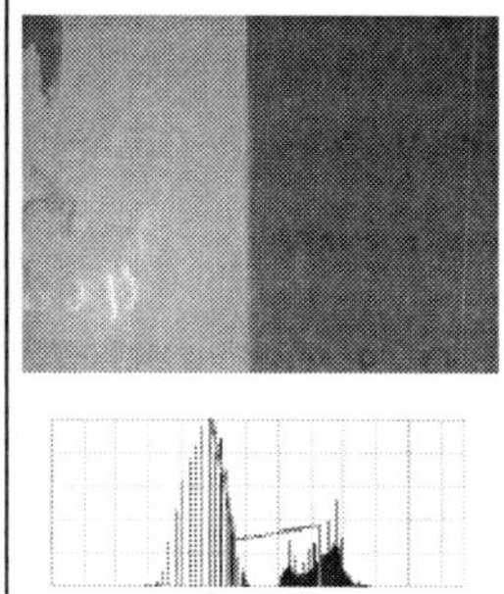 $b = 0.43$ $c = 0.26$

4. Results

The processing algorithm was tested in the welding environment. During the welding preparation (Fig. 13), the maximum contrast enhancement is provided. This offers a good view of the scene and the possibility to extract and recognize important elements like the inclination of the welding torch, the identification of the work piece and specific markers used for tracking to generate an Augmented Reality view, with additional visual elements.

Fig. 13 Image improvement during the welding preparation
a) input image; b) processed image

During welding (Fig.14), the position, shape and dimensions of the welding arc can be detected and the distance between the welding torch and the piece can be measured. Using the nonlinear transformation, image areas which are not illuminated by the arc (like the welding torch, surrounding of the welding arc) are enhanced with the possibility to track the inclination of the torch, and specific markers to detect the welding speed.

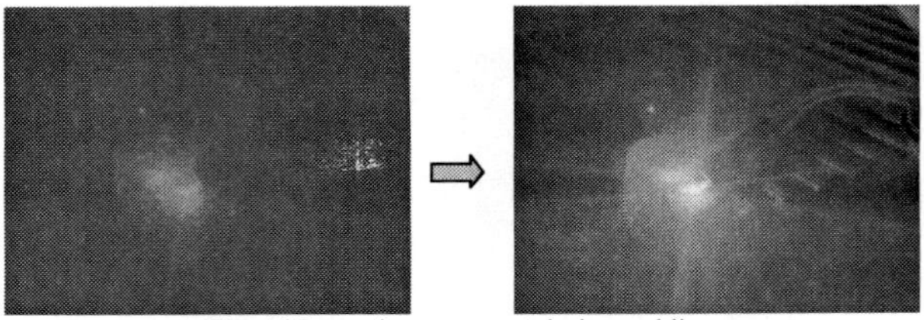

Fig. 14 Image improvement during welding
a) input image; b) processed image

If the illumination conditions change, namely when the weld is started or stopped the correction parameters are calculated and the image is adapted to the new conditions within milliseconds, offering a good protection for the user of the TEREBES system.

5. Conclusion

The paper describes the considerations and proposed solutions for the image quality improvement, defining the parameters to characterise the image brightness and contrast. The proposed linear corrections and nonlinear transformations are described, to be applied for different illumination conditions in the welding environment. The purpose is to offer more information about the welding scene, in order to extract additional visual data for an AR view for the operator.

References

[1] Gräser, A.; Tschirner, P.; Nordbruch, S.: *Virtual and Augmented Reality for Quality Improvement of Manual Welding*,11th International Conference on Computer Technology in Welding, Columbus, USA, 2001

[2] Tschirner, P.; Hillers, B.; Gräser, A.: *A Concept for the Application of Augmented Reality in Manual Gas Metal Arc Welding*, ISMAR 2002 30.09.-01.10.2002, Darmstadt

[3] Nordbruch, S.; Tschirner, P.; Gräser, A.: *Analyse von HDRC-Bildern des Werkstoffübergangs des MSG-Schweißprozesses*, (*Analyse of HDRC Images of Metal Transition in MAG-Welding*) 22. DAGM-Symposium, Kiel, 2000

[4] Aiteanu, D.; Gräser, A.: *Image pre-processing function using a non-linear Bezier transformation in observation of welding processes*, Third Conference on Mechanical Engineering – GEPESZET 2002, Budapest, Hungary 2002, p.462-466.

Literature

- *[5] Hypermedia Image Processing Reference*, Department of Artificial Intelligence in the University of Edinburgh, 2000
- *Computer Vision and Image Processing* – Chapmann and Hall, 1990
- *Computer Vision. Theory and Industrial Applications* – Carme Torras, 1992
- *Computer Vision. Principles* – Rangachar Kasturi, Ramesh Jain, 1994
- *Digital Image Analysis and Image Processing*, W.J. Hosack, University of Edinburgh, 1998
- *Applied Pattern Recognition*, Dietrich Paulus, Joachim Hornegger, 3rd Edition 2001

TEREBES – Enhancement of Visual Information in the Manual Welding Process

B. HILLERS, D. AITEANU, P. TSCHIRNER, A. GRÄSER

Institute of Automation, University of Bremen
Emails: hillers@, aiteanu@, tschirner@, ag@iat.uni-bremen.de

Keywords: *manual arc welding, augmented reality, HDRC camera, head-mounted-display, image processing, on-screen remote control, online documentation.*

Abstract

The manual welding process leads to very high demands for the welder, mainly because of high radiation and bad visibility of the welding scene due to the arc. In addition, the protection mask decreases the angle of view on the scene further. These hindrances may lead to a low seam quality, tedious working conditions and high post-processing cost if a seam of bad quality has to be reworked. An augmented reality system using a new developed welding helmet may improve the view of the welder. The scene is acquired by a stereoscopic High Dynamic Range CMOS (HDRC) camera system. The HDRC cameras provide a wide nonlinear dynamic range of $1{:}10^6$ concerning the light sensitivity. By image processing, the scene will be enhanced and displayed on a video see through Head Mounted Display (HMD). Besides the better view onto the welding process, the welder can operate an "on- screen" remote control of the welding machine to change the machine parameter. During the welding process, the inclusion of additional online process data is possible to augment the visual information. Another focus will be the opportunity to improve the quality management. Nowadays the quality of security relevant welding seams is tested with non-destructive material examination in a post-process. These methods like X-ray images or ultrasonic pulse echoing are costly and may disturb the work routine of the building site. Using TEREBES a documentation may be produced by storing the image together with electrical data. The documentation of welding data and welding images provide additional information like sputtering or idle-times, to mention only two. With an augmented reality helmet the welder will be able to produce seams of higher quality and the inspection cost will decrease. The article discusses the status and research of the helmet in detail.

1. Introduction

The purpose of the welding helmet is to improve the welding conditions and to verify the weld quality by documenting the seam producing process. The working conditions in the welding environment are difficult and demand for all the users abilities and skills. The main difficulties for the welder during manual welding and useful tasks of TEREBES has been analyzed together with welder in a shipyard because of the high requirements in domains like ship building [1]. The solutions presented in the following sections are based on that analysis.

During the welding process, the welder has no relevant information about the quality of the welding seam which he produces. The quality depends on the welders experience, based on the knowledge about the correct position and inclination of the torch, the correct distance to the piece and the suitable welding speed. To certify the quality of welding seams, high cost X-ray equipment is used. These complicated devices require specialized personal, and their use is time consuming.

The high quality requirements for welding seams in order to provide resistance and isolation, make necessary the removal of the whole produced welding seam, if only a small segment is defective. The removal and the production of a new welding seam implies extra human and material costs. If a welding system may recognize a defective area during welding, the producing process could be stopped immediately, saving time and resources.

When the necessity to change the welding machine settings occurs (by changing of welding conditions, i.e. the wire feed speed, the welding arc dynamic or voltage correction, the welding type etc), the user with a traditional helmet has to remove the welding helmet and to leave the working place to access the welding machine's panel. The implemented system offer the possibility to access the welding machine with the help of an "on-screen" user interface, without removing the helmet or leaving the place.

2. Proposed Solutions

The technical feasibility and the user acceptance have been realistically analysed to provide a concept with the possibility to certify the welding quality (by online monitoring and the quality documentation) and to improve the working conditions. An improved view to the welding scene is provided and direct access to the welding machine is offered. A quality supervisor or a learning assistant and additional constructive data round off the whole concept.

2.1 Improvement of working conditions

The enhancement of the image quality has been researched, to offer a better view than a regular welding helmet provides [2]. By using of video cameras with a high dynamic response, and by compensating the severe light difference before and during the welding process by adaptive image processing procedures, the objective is fulfilled (Fig.1).

a)

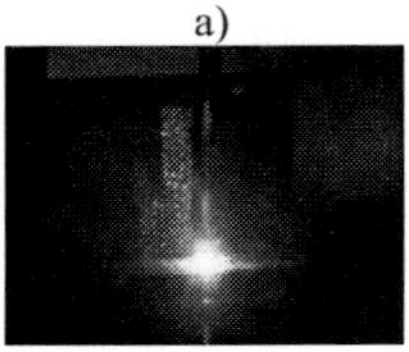

b)

Fig.1. Visual improvement using the AR system

a) the view of welder using regular welding helmet; b) the image using the AR system with image processing

To change the settings of the welding machine, an "on-screen" remote control, has been designed as a graphical interface, which can be blended on the border (bottom and right) of the displayable area (Fig.2).

Fig.2. Accessing the user menu

To select any available option, an input device is fixed on user's belt and can be easily operated with one hand and even with rough welding protection gloves. The device returns three basic functions (up, down, select) to navigate and make selections.

2.2 Image Improvement

2.2.1 Region independent improvement

The main advantage of using the HDRC cameras is the possibility to observe the welding arc and the environment of the welding scene at the same time. This is possible due the high dynamic characteristic of the CMOS camera chipset. As a disadvantage of this technique, the contrast in the images is very low, because of mapping the high dynamic range to a eight bit grey scale image (Fig. 3).

a) *b)*

Fig.3. The images provided by the HDRC camera; *a) before welding; b) during welding process*

The images has to be improved to provide the possibility to extract additional visual data. If the images are processed with the usual linear contrast enhancement, the results are satisfactory only for the images acquired before welding. In images acquired during the welding process, the information about the welding arc is destroyed by saturation (Fig.4).

a) *b)*

Fig.4. Enhancement of contrast using linear contrast stretching; *a) successfully improvement for images before welding; b) limited success for images during welding*

An new algorithm has been developed. In a first processing step, the brightness and the contrast are adjusted, considering that the video information in the brighter area has to be preserved. To accomplish this, the brightness and contrast correction are calculated in order to obtain in the processed image only a certain percentage of saturated pixels (pixels having the maximum admitted value for the light intensity). In this manner, the welding arc image is preserved, while the contrast for the images before welding are effectively enhanced.

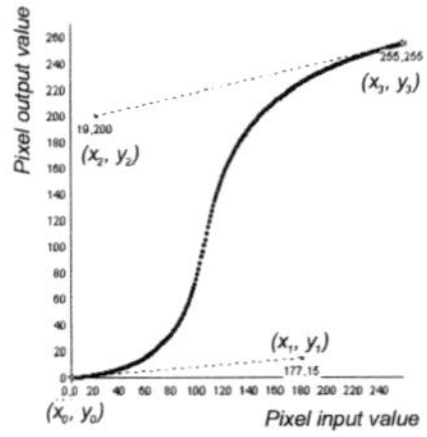

Fig.5. Sample of Bezier transformation curve

In a second step, a non-linear transformation is applied to the images containing the welding arc. The transformation is based on the Bezier curve properties (Fig.5) and provides the possibility to enhance the image contrast with different effects in the different image areas. By using of this procedure, the welding arc area is preserved, while the contrast in the environment is enhanced. The parameters of the transformation curve are dynamically determined, depending on the characteristics of the input image, to offer the optimal improvement [3](Fig.6).

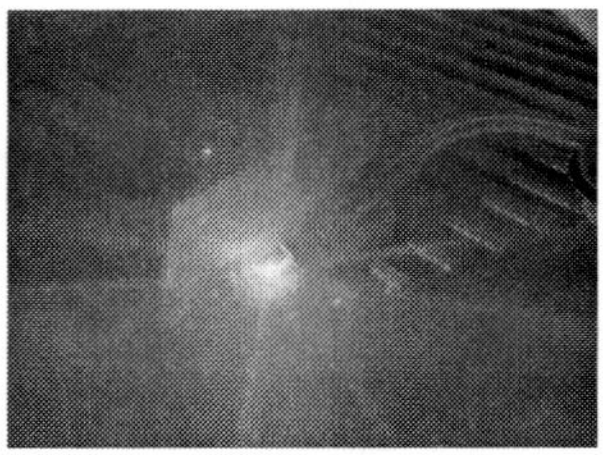

Fig. 6. The processed image during welding

2.2.2 Region dependent improvement

The preceeding chapter describe the nonlinear but region independent enhancement of image contrast and brightness. To provide a further improvement of the image quality an image region dependent technique is suggested. The image contrast is adapted to the different image areas, so that the resulting image contains all image areas with sufficient contrast. Standard image processing techniques cannot cope with the extreme brightness conditions and the characteristics of the camera. Objects in far distance of the arc may differ only by few gray levels from the background. In these image areas a contrast enhancement is necessary. Objects close to the arc are clearly visible against the background. In these areas a contrast enhancement can lead to a degradation of the image quality (e.g. by exceeding the range of the gray level images). Therefore the image processing procedure needs to be robust against varying image contents. Contrast and brightness has to be adapted to the different image areas, so that the resulting image contains all image areas with sufficient contrast. An adaptive procedure has been developed that "recognizes" the high-contrast areas, which have to be maintained, and the low-contrast areas which have to be improved.

The improvement of the image quality is based on the method "Contrast Enhancement Using the Laplacian-of-a-Gaussian filter", which was modified for the present application [4]. With Neycenssac's method the enhanced image $g(x,y)$ is obtained by the convolution of the original image $f(x,y)$ with a LoG filter $LoG(x,y)$, multiplication of the resulting image $g'(x,y)$ by a contrast factor β and subtraction from the original image $f(x,y)$:

$$g(x,y) = f(x,y) - \beta \cdot f(x,y) * LoG(x,y) \quad (1)$$

For the implementation of an adaptive contrast enhancement the procedure was modified by adapting the contrast factor β to the image content. To generate a priority function for β a gradient operator is applied to the original image in order to determine the contrast of the image areas. The resulting gradient image $f_{Sob}(x,y)$ is low-pass filtered to create a continuous transition between the areas. Within high-contrast areas, the low-pass-filtered gradient image $f_{Sob,TP}(x,y)$ contains high values, and β must be small in order to maintain the contrast. Within low-contrast areas, the low-pass-filtered gradient image $f_{Sob,TP}(x,y)$ contains low values, and β must be high in order to enhance the contrast. The following relationship for β was experimentally determined:

$$\beta = 200 - 40 \cdot \ln[f_{Sob,TP}(x,y)] \quad (2)$$

Figure 7 shows the concept of the procedure. An example for a result of image processing with uniform contrast in all image areas is shown in figure 8.

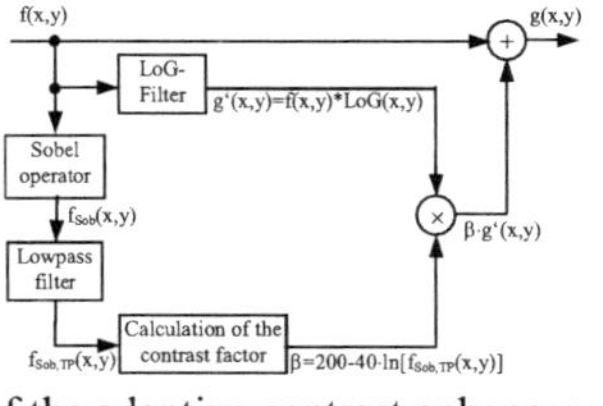

Fig. 7. Flow of the adaptive contrast enhancement procedure

(a)

(b)

Fig. 8. (a) Original Image; (b) Result of adaptive contrast enhancement

2.3 Quality Management

To improve the quality of the manual welding is one of the systems ambitions. This application will improve the information environment of the welder before and during the welding process and will give the quality manager the access to online and offline information. The welder has an "on-screen" remote control of the welding machine by choosing the program for the specific welding, including detailed information about the

selected program. All these options are provided without lifting the welding helmet or changing the welders position. During welding, some information about the process may be displayed, as welding current and voltage. In this way the welder may react directly if the measurement of the current or voltage differs from pre-selected boundary values.
To check the seam quality radiographies or ultra sonic tests are made With TEREBES it is able to look directly into the welding process and it is possible to document the producing process of the welding seam itself. At first glance weld preparation, weld droplets, idle-periods and sputtering will be viewable. This in vivo documentation can be correlated to radiographic or ultrasonic tests and may replace some of these test in future

3. Implementation

In the following section the implementation of the mentioned requirement specs and the affiliated expected AR-features is discussed. The chapter is divided into three parts for the helmet and hardware and software realisation.

3.1 Helmet

The helmet has to protect the welder against the ultraviolet rays of the welding arc [5]. It has to be robust against mechanical shocks concerning the stability and functionality of the built-in HDRC-cameras and HMD Fig 9 and 10 show the present design of the helmet.

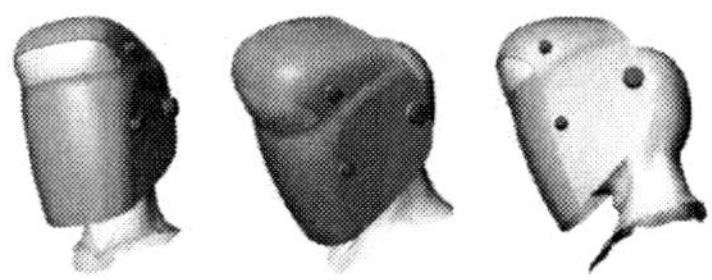

Fig.9. Design of the welding helmet

The HDRC Cameras are hidden behind a renewable screening grid, which shields the optics against smoke and welding sputter. The HDRC-cameras are built into the helmet above the eyes and in front of the users forehead. The experience which are made up to now, show that this position although not in the line of sight of the user has no bad influences to an ergonomic view on the HMD.

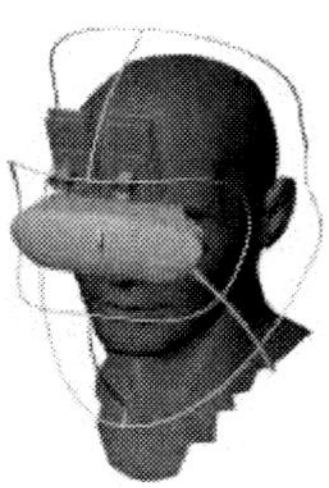

Fig.10. Built in HMD and HRDC cameras

The Head Mounted Display is installed directly into the helmet and can be adapted mechanically to the users needs. To choose an individual position of the HMD is a critical point in achieving a good view because of its resolution restriction and the low possible angle of view of the TFT-displays. Without a good adaptation of the HMD position to the users eyes no 3D impression is reached.

3.2 Hardware

The computer system consists of three embedded and a stationary PC to ensure the needed computing power (Fig.11).

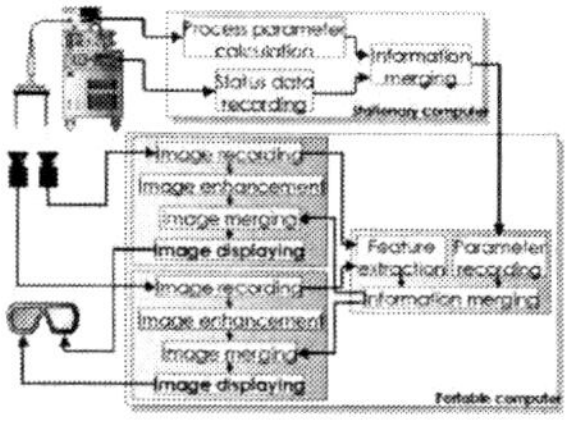

Fig.11. Hardware system [6]

As introduced, three embedded PCs equipped with Intel Pentium III / 800 MHz CPUs are interconnected by a LAN. Two of the three PCs serve a single HDRC-camera each and the accordant screen whereby the stereo camera system of the helmet is fully connected. The images provided by the left and right camera are directly and independently displayed on the left and right display of the HMD.

The third embedded PC will be used to extract information from the camera images and to provide an input device to control the application. The extraction of data from the HDRC images will be executed at a reduced frame-rate compared with the camera and HMD frame rate. Features like the distance of welding torch to the work piece or the welding torch angle may be extracted by this PC. As well the movement of the helmet by video based navigation using the information of the stereo camera system to know where to place virtual objects in the scene.

Using three embedded PCs it is necessary to distribute also the taks among the PCs. The concept of **o**bject **o**riented **p**rogramming (OOP) and the Common Object Request Architecture (CORBA [7]) is used for this task. CORBA is defined by the Object Management Group (OMG) and supplies access to created objects and is independent from the used domain, operating system and programming language. The computer are linked via a Fast Ethernet (100MBit) network which constitutes the limiting factor for the maximum frame rate of the 8 Bit greyscale images with 800 by 600 pixel resolution. With an upgrade to Gigabit Ethernet, this limitation will be overcome.

To control the Graphical User Interface (GUI) an input device, similar to a wheel mouse is used. The application is controlled by scrolling and pressing the wheel button. The data from the wheel is received by the third PC and synchronizes the menu display on the two camera PCs. The GUI will be described detailed later on in the software chapter.

3.3 Software

3.3.1 The User Menu

To access the additional data like the welding parameters, to set the welding machine configuration, or to make available the AR data, a graphical user interface has been developed, to be operated by the welder directly on the helmet (Fig. 2).

By pressing the key of the input device, two bars are blended on the border area of the image. At the right hand side, there is a vertical bar containing virtual buttons. The currently selected button is highlighted. To select the desired button, the input device wheel can be scrolled up

and down. By pressing the input device button the computer executes the routine assigned to the respective button.
Besides the basic options (i.e. to close the menu, or to confirm the changed made), one of the implemented functions is to display the welding parameters (welding voltage and welding current) (Fig. 12).

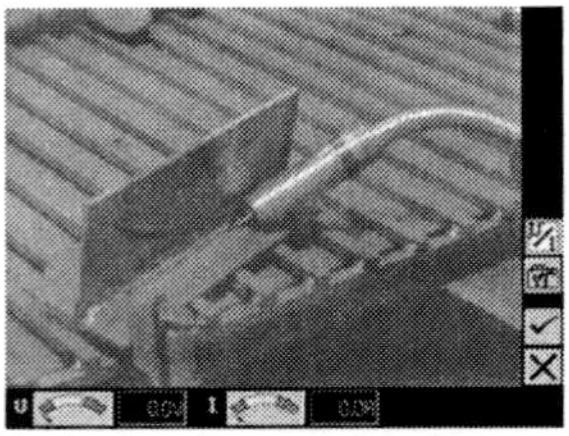

Fig.12. Displaying the actual welding parameters

The values are displayed as numerical values, as well as "analogue" indicators showing the tolerance limits on a green/red scale. In this manner, during the welding process, the welder has an indication about a possible welding error without specially focusing on the indicators. By pressing the button again, the display of welding parameter is turned off.

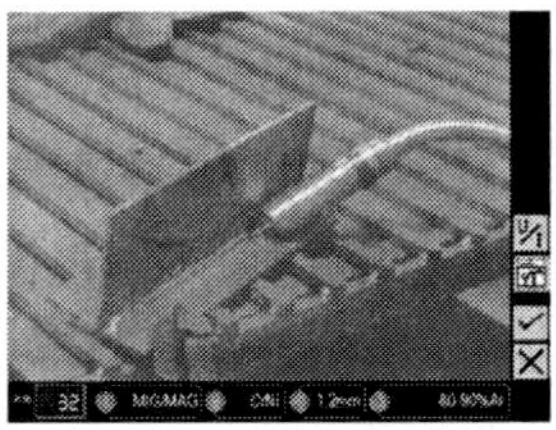

Fig.13. Setting the welding machine parameter

Another implemented option is the adjustment of the welding machine configuration, i.e. to set the assigned "job" for the actual welding specification:
weld type, material type, wire diameter, gas composition (Fig.13). If this option is selected, the scroll of the input device's button are assigned automatically to browse the list of possible jobs. The meaning of the selected job is displayed beside the job number. If a certain job is selected and no other command is given for a certain time (in this case, 2 seconds) the selected job is sent to the machine and a confirmation is displayed on the screen. If the input device's button is pressed sooner than 2 seconds, the computer understands that the change of parameter was cancelled, and keeps the previous configuration of the welding machine.
By using of these options, the user has direct access on the welding machine information. The computer analyses the welding parameter and gives an indication when the welding seam quality is critical and the process has to be interrupted, with the benefit of saving time and additional costs. Also, the user is able to change the machine settings, without removing the helmet or leave the work place. Together all this features create a friendly and ergonomic working environment.

3.3.2 The Welding Machine Commander

The program running on the stationary computer has two main functions. First, the computer provides the communication with the welding machine via the serial interface. One software module is coding/decoding the information to/from the welding machine, makes the data available through the communication server (on a CORBA platform) to be processed by the monitoring system on the portable computer or to be saved in the quality documentation. The server accepts as well some commands from the remote clients, concerning the settings of the machine.

An enlarged possibility to access the welding machine is provided by a dialog (Fig.14). Besides displaying of basic welding parameter (current, voltage), the wire-feed speed is displayed. The status of the communication with the welding machine is detected: initializing, communication running or no signal. There is also the possibility to change the length and the focus of the welding arc by changing the values for dynamic and voltage correction.

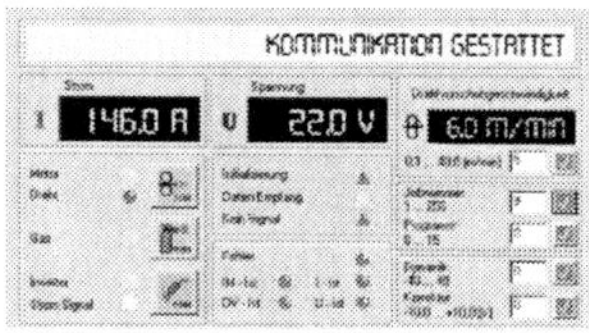

Fig. 14. The welding machine commander

Another option is to test the functionality of the welding machine (wire-feed test, gas test, weld test) and to report some detected errors. Some hardware detected errors, like current, voltage or wire-feed speed error, as well as general machine failure are reported.

4. Conclusions and Outlook

The possibility to implement some new concepts has been realistically analysed, concerning the technical feasibility and the user acceptance. The concept provides the possibility to certify the welding quality (by online monitoring) and to improve the working conditions, by offering direct access to the welding machine and a quality supervisor.

[1] Park, M. ; Canales, C. ; Schmidt, L. : User Interface for a Wearable Augmented Reality Welding System, ERCIM Research Report, Workshop Adjunct Proceedings, N.Carbonell, C. Stephanidis, Paris, France 2002, S. 167 - 174.

[2] D. Aiteanu, D.; A. Gräser: Image pre-processing function using a non-linear Bezier transformation in observation of welding processes, Third Conference on Mechanical Engineering – GEPESZET 2002, Budapest, Hungary 2002, p.462-466.

[3] D. Aiteanu, A. Gräser: Image processing function for compensation of low-contrast images acquired during welding process, 24. Kolloquium der Automatisierungstechnik, 2002, Salzhausen, Germany

[4] P. Tschirner: Visuelle Analyse von HDRC-Kamerabildern des Werkstoffübergangs des MSG-Schweißprozesses, Diplomarbeit, Institut für Automatisierungstechnik, Universität Bremen, 1999.

[5] P. Tschirner, B. Hillers, A. Gräser: A Concept for the Application of Augmented Reality in Manual Gas Metal Arc Welding, ISMAR 2002 30.09.-01.10.2002, Darmstadt

[6] P. Tschirner, S. Nordbruch, Axel Gräser: Quality Improvement of Manual Welds by Virtual and Augmented Reality, International Sheet Metal Welding Conference X
14.-17.05.2002, Detroit, Michigan, USA, Paper 4-6

[7] M. Henning, S. Vinoski: Advanced CORBA Programming with C++, Addison Wesley 1999, ISBN 0201379279